DICTIONARY OF BOTANY

by

PARAMESHWAR HEGDE

Gaurav Book Centre Pvt. Ltd.

4832/24, Prahlad Lane, S-207 Ansari Raod, Daryaganj, New Delhi-110002

Published by
Gaurav Book Centre Pvt. Ltd.
4832/24, Prahlad Lane, S-207 Ansari Raod, Daryaganj,
New Delhi-110002
Tel.: 011-43570976, 23278261
Email-gauravbookcentre@gmail.com

First Edition : 2013

ISBN : 978-93-80317- 30-4

Price : 200/-

Laser Typesetting by : Vinod Bhardwaj

Printed at: Vikas Computers, Delhi

PRINTED IN INDIA

Published by Gaurav Book Centre Pvt Ltd,New Delhi-110002

PREFACE

Botany, plant science(s), or plant biology , a discipline of biology, is the science of plant life. Traditionally, the science included the study of fungi, algae, and viruses. A person engaged in the study of botany is called a botanist.

Botany covers a wide range of scientific disciplines including structure, growth, reproduction, metabolism, development, diseases, chemical properties, and evolutionary relationships among taxonomic groups. Botany began with early human efforts to identify edible, medicinal and poisonous plants, making it one of the oldest branches of science. Nowadays, botanists study about 400,000 species of living organisms.

The study of plants is vital because they are a fundamental part of life on Earth, which generates the oxygen and food that allow humans and other life forms to exist.

The present book covers thousands of botanical words and their meanigs.I hope this book would be very useful for students ,Teachers,lecturers, andalso for common people.I would like to express my sincere thanks to the publishers for their continued interest in bringingout this edition.

Parameshwar Hegde
M.Sc.,M.A.

A

ABAXIAL. Abaxial means being located on the side away from the axis. The abaxial surface of a leaf is its underside.

ABSCISIC ACID. Abscisic acid is a plant hormone that inhibits growth, causes the abscission of leaves, induces dormancy, closes stomata, and triggers other phenomena in response to adverse conditions.

ABSCISSION. Abscission is the normal separation of a leaf, fruit, or flower from a plant. Abscisic acid is the plant hormone involved in abscission.

ABSCISSION ZONE. The abscission zone is the area at the base of leaf's petiole, a fruit stalk, or a branch in which the separation (abscission) layer develops. The disintegration of this layer causes a leaf, fruit, or flower to fall from a plant. Abscisic acid is the plant hormone involved in this process.

ACICULAR. Acicular means needle-like (for example, pine needles are acicular).

ACID RAIN. Acid rain is polluted and harmful to the environment. Acid rain has a low pH. Acid rain may have been a component of the K-T extinction.

ACHENE.

An achene is a dry, indehiscent one-seeded fruit with a leathery pericarp that is easily separated from the seed coat (for example, sunflower).

ADAPTIVE RADIATION. Adaptive radiation is the diversification of a species as it adapts to different ecological niches. If successful, the species becomes specialized for the new environments (the mechanism being natural selection), and they eventually evolve into different species.

ADAXIAL. Adaxial means being located on the side towards the axis. The adaxial surface of a leaf is the upper side.

ADENOSINE TRIPOSPHATE. ATP (short for adenosine triphosphate) is a nucleotide that has a ribose sugar and three phosphate groups. ATP is a high-energy molecule used for energy storage by organisms. In plant cells, ATP is produced in the cristae of mitochondria and chloroplasts.

ADVENTITIOUS. Adventitious organs are those organs that grow from an unusual part of the plant. For example, fibrous adventitious roots grow from the trunk above the ground instead of starting underground.

AERIAL ROOT. An aerial root is a plant's root that is produced above the ground.

AGE. An age is a unit of geological time which is distinguished by some feature (like an Ice Age). An age is shorter than epoch, usually lasting from a few millions of years to about a hundred million years.

AGRICULTURE. Agriculture is the science of farming, including growing plants and raising animals.

AGROFORESTRY. Agro-forestry is a land use system in which woody perennials are grown with agricultural crops (together with other land uses, like animal production).

AGROLOGY. Agrology is a branch of soil science that studies the soil used in producing crops.

AGROSTOLOGY Agrostology is a branch of botany that studies grasses.

AIR SPACE. Air space is the intercellular gaps within the spongy mesophyll of leaves. These gaps are filled with gas that the plant uses (carbon dioxide - CO_2) and gases that the plant is expelling (oxygen - O_2, and water vapor).

AIR PLANT. Air plants (also called epiphytes) are plants that live attached to a plant (or other structure like a telephone pole or a building) and not in the ground). Epiphytes include many orchids and bromeliads. Epiphytes are not parasites; they get water and nutrients from the air (and not from their host).

ALGAE. Algae are simple photosynthetic organisms that belong to the kingdom Protista. Most algae are aquatic; seaweeds are algae. Some algae are unicellular while others are multicellular

ALIEN. An alien is a plant that is not native to a place; it came from another place.

ALOE

Aloes are succulent, clumping plants with fleshy, toothed (non-fibrous) leaves. THere are about 300 species of aloe; they live in warm, dry habitats and most originated in northern Africa. Aloe vera is a popular plant whose gel-like sap is used as a medicinal salve; it is originally from northern Africa, the Cape Verde Islands, and the Canary Islands. Aloe vera has gray-green leaves and tubular yellow flowers that grow on a stalk that is up to 3 feet tall. Classification: Family Liliceae (lilies). Genua Aloe, Species - about 300 including *A. vera*, *A. ferox*, etc.

ALTERNATE.

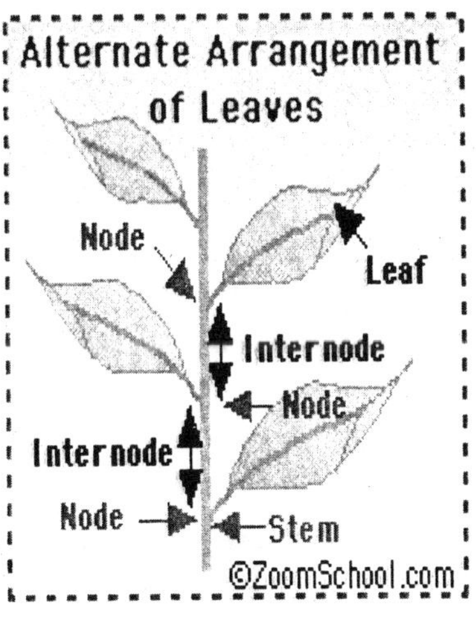

An alternate pattern of leaves or buds is a pattern in which there is one leaf (or bud) per node, and on the opposite side of the stem (not in pairs).

ALVAREZ THEORY OF EXTINCTION. This theory is

that a large asteroid, meteor, or comet hit the Earth 65 million years ago, causing huge atmospheric and geologic disruptions, leading to a mass extinction which killed the dinosaurs and many other plant and animal species.

AMBER.

Amber is a yellowish, fossilized tree resin (from conifers) that sometimes contains bits of trapped matter.

AMYLOPLAST. A organelle (with double membranes) in some plant cells that stores starch. Amyloplasts are found in starchy plants like tubers and fruitsr.

ANGIOSPERM.

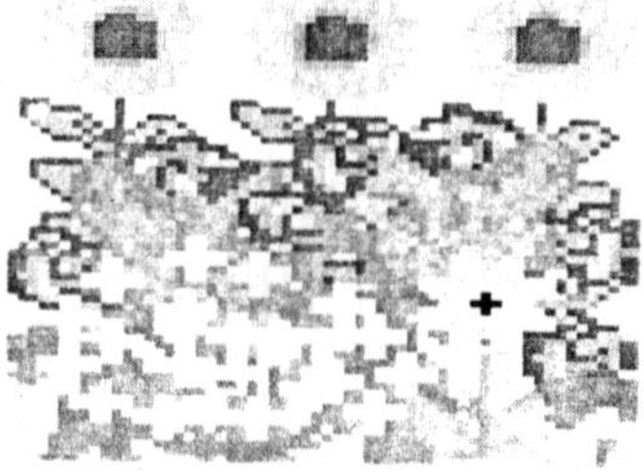

(pronounced AN-jee-oh-sperm) Angiosperms (meaning "covered seed") are flowering plants. They produce seeds enclosed in fruit (an ovary). They are the dominant type of plant today; there are over 250,000 species. Their flowers are used in reproduction. Angiosperms evolved about 145 million years ago, during the late Jurassic period, and were eaten by dinosaurs. They became the dominant land plants about 100 million years ago (edging out conifers, a type of gymnosperm). Angiosperms are divided into the monocots (like corn) and dicots (like beans).

ANNUAL. An annual is a plant that goes through its entire life cycle within a year. It grows from a seed, matures, produces seed, and dies within a year.

ANNUAL RING. Annual rings are concentric circles that appear on tree trunk cross-sections that mark the end of a growing season. These rings show whether the tree grew a lot or a little that year.

ANTHER.

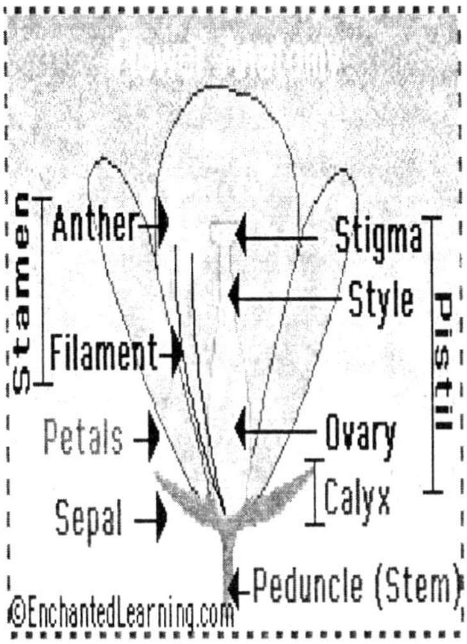

The anther is the tip of a flower's stamen. The anther contains the pollen.

ANTHOPHYTA Anthophyta are flowering plants, the largest group of plants (which includes the grasses). The flowers are used in reproduction. They evolved during the Cretaceous period.

APICAL DOMINANCE. Apical dominance is the phenomenon in which a terminal (end) bud inhibits the development of lateral (side) buds.

APICAL MERISTEM The apical meristem consists of meristematic cells located at the tip (apex) of a root or shoot.

APOGEOTROPIC ROOTS. Apogeotropic roots are roots that grow upwards to the soil surface (other roots grow downwards), emerging from the soil and growing upwards.

APOMIXIS. Apomixis is a type of reproduction in which a plant produces seeds without fertilization.The sego palm has apogeotropic roots, as do cycads.

APOMORPHY. An apomorphy is a new genetic characteristic common to a clade. Feathers are an apomorphy for birds.

APPLESEED, JOHNNY.

Johnny Appleseed was a man who spread apple trees through the USA. His real name was John Chapman, but he was called Johnny Appleseed because of his love for growing apple trees.

AQUATIC. Aquatic organisms are those found in water. Many plants are aquatic, including seaweeds.

ARABLE. Arable land is suitable for growing crop plants.

ARBOREAL. Arboreal means living in trees. Many animals are arboreal, including the sugar glider.

ARBORETUM. An arboretum is a park or garden where trees and shrubs are grown for educational and/or scientific uses.

ARCUATE. Leaves with arcuate venation have veins that are curve towards the apex (tip).

AREOLE. Areoles are circular clusters of spines on a cactus. Flowers bud at an areole and new stems branch from an areole.

ARID. An arid area is dry and hot, with little rainfall and few plants.

ASTEROID. An asteroid is a large rock or small planet orbiting the Sun. Most asteroids lie in a belt between Mars and Jupiter. An asteroid impact with the Earth may have caused the K-T mass extinction.

ATMOSPHERE.

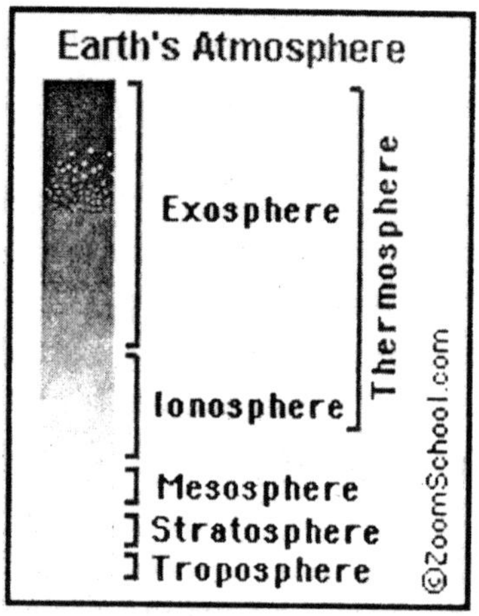

The atmosphere is the mixture of gases that surrounds the Earth. The Earth's atmosphere is mostly nitrogen.

ATOM.

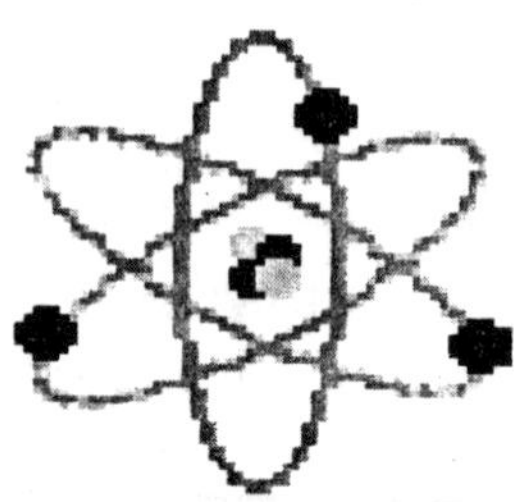

Everything is made up of tiny atoms. An atom is the smallest part of an element that has the properties of that element.

ATP. ATP is short for adenosine triphosphate; it is a nucleotide that has a ribose sugar and three phosphate groups. ATP is a high-energy molecule used for energy storage by organisms. In plant cells, ATP is produced in the cristae of mitochondria and chloroplasts.

AUTOTROPH.

(pronounced AW-toh-trofe) An autotroph (or producer) is an organism that makes its own food from light energy or chemical energy without eating. Most green plants, many protists (one-celled organisms like slime molds) and most bacteria are autotrophs. Autotrophs are the base of the food

AUXIN. Auxins are growth hormones found in plants. Auxins induce phototropism, apical dominance, cell elongation and many other reactions.

AWN. The awn is a bristle-like extension of a plant near its tip.

AXIL

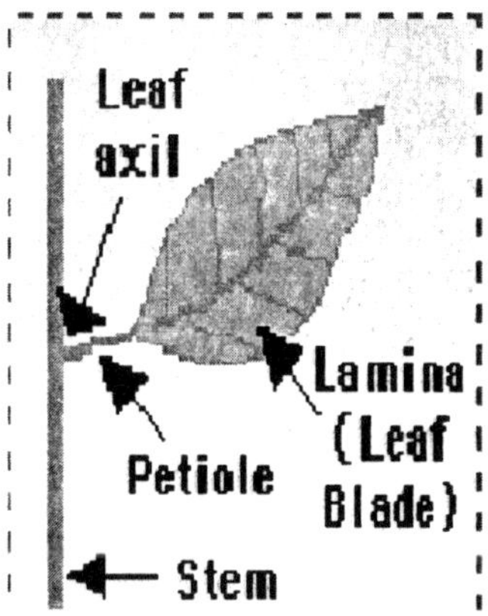

The axil of a plant is the angle between the upper side of the stem and a leaf, branch, or petiole. In flowering plants, the bud develop in the axil of a leaf

AXILLARY BUD. The axillary bud is a bud that develops in the axil (the angle between the stem and the leaf) of a plant.

AXIS. The axis of a plant runs through the middle of it, e.g., the stem of a plant or the rachis of a compound leaf.

B

BACKGROUND EXTINCTIONS. Background extinctions are those extinctions that occur continually throughout time. These extinctions are caused by small changes in climate or habitat, depleted resources, competition, and other changes that require adaptation and flexibility. Most extinctions (perhaps up to 95 per cent of all extinctions) occur as background extinctions.

BACTERIUM.

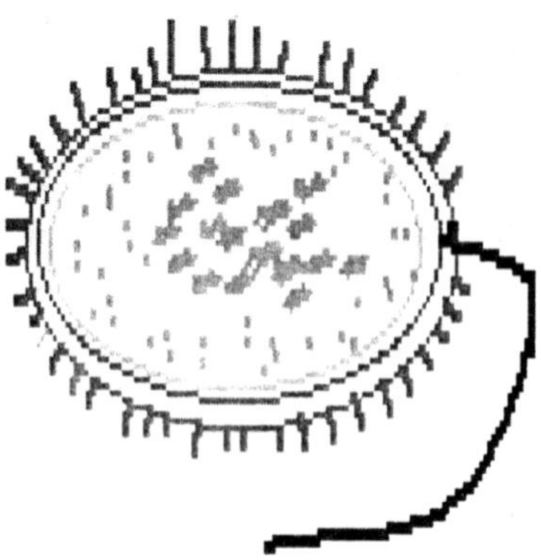

A bacterium (plural bacteria) is a microscopic, prokaryotic, single-celled organism with a cell wall. Bacteria reproduce by binary fission. Bacteria are one ot the earliest forms of life.

Some bacteria (rhizobia) are nitrogen-fixing; they transform nitrogen gas in the atmopshere into a form that can be used by plants.

BADLANDS. Badlands are barren, severely eroded places on Earth where the soft rock layers are sculpted into beautiful forms. These exposed rock layers are often wonderful places in which to find fossils. They're called badlands because the land is useless for farming and many other human purposes.

BAMBOO

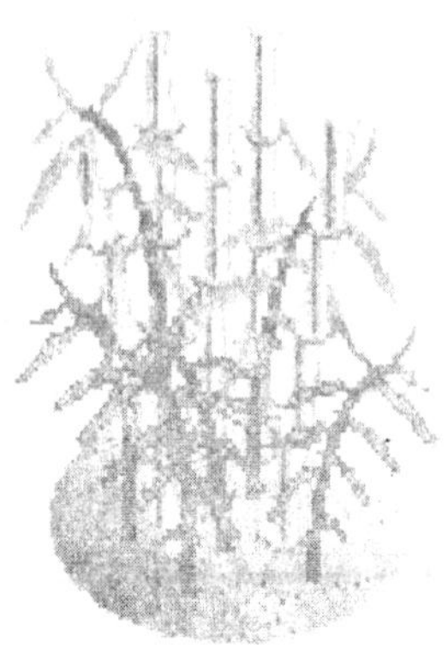

Bamboo is a fast-growing monocot grass.

BANANA. The banana tree is an herbaceous plant (because there is no woody tissue in the stem), it is also considered a fruit (because the seeds of the plant are contained within the banana).

BARK. The bark is the outer covering of the trunk, branches, and roots of trees.

BARLEY. Barley is a cereal grain related to wheat, oats, and rice. Barley seeds (the part of the plant that is highest in nutrition) grow in spikes at the tips of the stems - it resembles wheat. This important crop is grown in temperate areas and is used as animal feed, for malt (used in making beer, malted milk, and food flavoring), and for human consumption (used as a flour, a thickener, a cereal, in soups, etc.). Barley was cultivated thousands of years ago; it was grown by the Egyptians from 5,000 to 7,000 years ago. The countries that produce the most barley are Russia, Canada, and Germany. Classification: family Poaceae (Gramineae) (grasses), genus Hordeum, species H. vulgare.

BENNETTITALEANS

Bennettitaleans, also called cycadeoids, are primitive plants (gymnosperms) that resemble cycads (but they are not cycads), but have different methods of reproduction. Bennettitaleans lived throughout the Mesozoic Era. Examples of bennettitaleans include Williamsonia (Jurassic through end Cretaceous), Williamsoniella (Jurassic through end Cretaceous), and Zamites (Triassic).

BERRY. A berry is a small, juicy, fleshy, stoneless fruit that contains one or many seeds. This simple fruit has a pulpy pericarp surrounding the seed. Some berries include the gooseberry, tomato, currant, and grape.

BIENNIAL .Biennial (menaing "two years") means that it takes two years to complete the full life cycle.

BILATERAL .Bilateral means two sides.

SYMMETRY .Bilateral symmetry (also called symmetry across an axis) is when one side of an object is the mirror image of its other half - i.e., one half has the same shape and size as the other half (for example, most leaves are bilaterally symmetrical).

BILLION. .A billion is a thousand million. Multicellular life evolved on Earth about a billion years ago.

BINOMIAL NOMENCLATURE. Binomial nomenclature is a system developed by Linnaeus for giving organisms scientific names in which each organism has a genus name (always capitalized) and a species name (not capitalized). For example, the Venus' flytrap is *Dionaea muscipula*, and people are *Homo sapiens*.

BIODIVERSITY .Biodiversity is the abundance of different plant and animal species found in an environment.

BIOLOGICAL MAGNIFICATION .Biological magnification is the phenomenon in which toxins (poisons) are more and more concentrated in living organisms that are higher up in trophic levels of the food web. For example, if a small amount of toxins is in plants, the animal that eat those plants have a higher concentration of the toxins, and the meat-eaters that eat those plant-eaters have even higher levels of the toxin. The toxins are from pesticides (bug killing chemicals), herbicides (weed-killing chemicals) and waste materials.

BIOMASS. A biomass is the total amount of living material in an area. It is calculated by adding up the weights of all of the organisms.

BIRCH. Birch (genus Betula, many species) are broadleafed, deciduous trees and shrubs with paper-like bark.

BLADDER. A bladder is a small, air filled sac. Some plants, like bladderwort (a seaweed), have air bladders.

BLADE. A blade is a narrow, flat leaf.

BOTANIST. A botanist is a scientist who studies plants.

BOTANY. Botany is the scientific study of plants.

BRACT. A bract is a reduced, leaf-like structure that is associated with a flower or a cone.

BRANCH. A branch is a part of a tree or shrub that grows from the trunk or stem.

BROMELIAD. Bromeliads are a group of plants that have stiff, waxy leaves that form a cup-shaped body. This "cup" catches and retains water during wet weather, and the plant uses this water to live through dry spells. Most bromeliads are xerophytes (able to tolerate a dry environment) and epiphytes (living attached to another plant and not rooted in the ground). Classification: Division Magnoliophyta (Angioperms), Class Liliopsida (monocots), Subclass Zingiberidae, Order Bromeliales, Family Bromeliaceae (Bromeliads). Bromeliads were named for the Swedish botanist Olaus Bromelius (1639-1705).

BROWSER. A browser is an animal that eats tall foliage (leaves). Many sauropod dinosaurs, like Brachiosaurus and Ultrasauros, were browsers.

BRYOPHYTES. Bryophytes (Phylum Bryophtya) include the hornworts (Class Anthocerotae), liverworts (Class Hepaticae), and mosses (Class Musci). These low-growing plants do not have true roots, leaves, or stems - they also lack a vascular system (the internal tubes that transport food and water in more advanced plants). Bryophytes probably evolved from green algae; fossils of bryophytes are rare.

BUD. A bud is a small, developing part of a plant that will grow into a flower, a new leaf or a stem.

BUD SCALE. A bud scale is a modified leaf (or similar structure) that covers and protects the bud.

BULB.

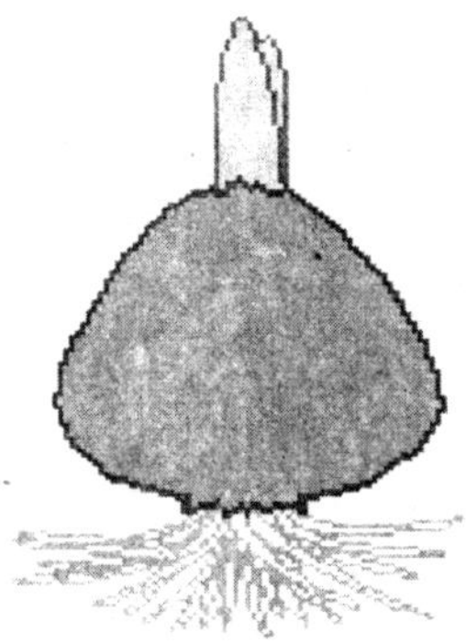

A bulb is an underground stem, usually globular, that has fleshy leaves emerging from the top and roots emerging from the bottom. The fleshy leaves store food. Examples include the tulip, narcissus, and onion.

BULBEL. A bulbel (also called a bulbet) is a small bulb that grows from another bulb. This is an example of vegetative proagation.

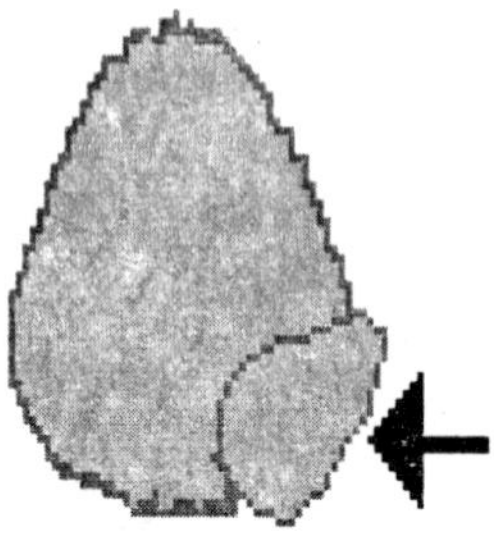

BULBLET. A bulbet (also called a bulbel) is a small bulb that grows from another bulb. This is an example of vegetative proagation.

BURBANK, LUTHER.

Luther Burbank (1849-1926) was an American plant breeder who developed over 800 new strains of plants, including many popular varieties of potato, plums, prunes, berries, trees, and flowers. One of his greatest inventions was the Russet Burbank potato (also called the Idaho potato), which he developed in 1871. This blight-resistant potato helped

Ireland recover from its devastating potato famine of 1840-60. Burbank also developed the Flaming Gold nectarine, the Santa Rosa plum, and the Shasta daisy. Burbank was raised on a farm and only went to elementary school; he was self-educated. Burbank applied the works of Charles Darwin to plants. Of Darwin's *The Variation of Animals and Plants under Domestication*, Burbank said, "It opened up a new world to me."

BURGESS SHALE. The Burgess shale is an incredibly fossil-rich area in the Canadian Rocky Mountains (in British Columbia). This Lagerstatten (a geological fossil deposit rich with varied, well-preserved fossils) is replete with fossils from the Cambrian Period, roughly 500 million years old. The Burgess shale was discovered in 1909 by Charles Doolittle Wolcott, who was the Secretary of the Smithsonian Institution in Washington, D.C. at the time. Fossils from this area include early representatives of most modern groups, including plants, worms, sponges, shrimp-like crustaceans, and jellyfish.

BUTTRESSES. Buttresses are tree trunk supports that help hold up tall trees in rainforests. They are needed to stabilize the tree because the soil is shallow (only a few inches deep) and the tree roots do not penetrate very deeply into the earth

BYA. "bya" is an abbreviation for billions of years ago.

C

C3 PLANT. A C3 plant is one that produces phosphoglyceric acid, (a molecule that has three carbon atoms) as a stable intermediary in the first step in photosynthesis (the Calvin Cycle). Most plants on Earth (over 95 percent) are C3 plants.

C4 PLANT. A C4 plant is one that produces oxaloacetic acid (a molecule that has four carbon atoms) as a stable intermediary in the first step in photosynthesis. Very few plants on Earth (less than 1 percent) are C4 plants (including corn and sugarcane).

Photorespiration in C4 plants is more efficient in strong light. The processes in C4 biochemistry were studied by M. D. Hatch and C. R. Slack.

CACAO. The cacao plant (*Theobroma cacao*) is a evergreen flowering tree native to wet, warm forests of South and Central America. This tree grows to 40 feet (12 m) tall. After flowering, 10 to 14-inch long red fruit pods develop. In each pod are almond-shaped cacao beans and pulp. Chocolate is made from the beans in the pods of the cacao plant.

CACTUS. A cactus (the plural is cacti) is a succulent plant that can live in dry areas. It has a structure that minimizes water loss. The stems are photosynthetic, green, and fleshy. The leaves are reduced to spines or are absent. Classification: Division Magnoliophyta (angioperms), Class Magnoliopsida (dicots), Subclass Caryophyllidae, Order Caryophyllales, Family Cactaceae (Cactus).

CALIFORNIA POPPY.

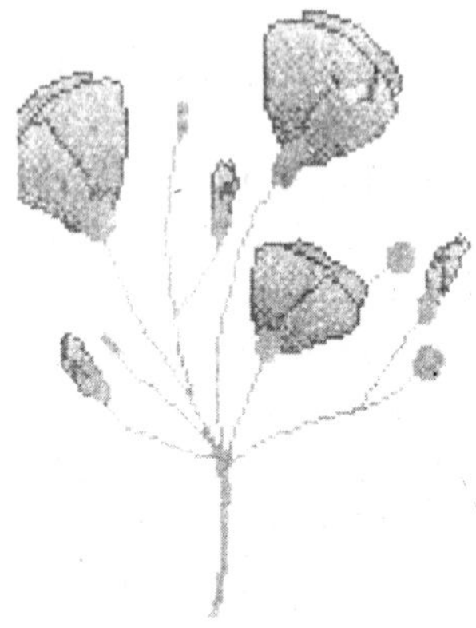

A golden poppy (*Eschscholzia californica*) from western North America. It has finely-divided foliage and cup-shaped flowers.

CALVIN CYCLE. The second stage in the process of phtosynthesis is called the Calvin Cycle (it is also called the Calvin-Benson Cycle or the Carbon Fixation Cycle. In the Calvin Cycle, carbon molecules from carbon dioxide, CO_2, are fixed into the sugar glucose, $(C_6H_{12}O_2)$ (in six repeats of the cycle). The Calvin Cycle takes place in the stroma of eucaryotic chloroplasts. The major enzyme that mediates the Calvin Cycle is Rubisco (ribulose-1-

5-biphosphate carboxylase). The Calvin Cycle was first investigated in the late 1940s and early 1950s by the Nobel Prize winning chemist Melvin Calvin (1911-1997).

CALYX . The calyx is the sepals of a flower.

CAMBIUM. Cambium is a layer of dividing cells found in the stems of plants. The cambium forms the specialized xylem and phloem cells and causes the stem to increase in thickness.

CANOPY. The canopy consists of the upper parts of the trees of a rainforest (about 65 to 130 feet or 20 to 40 m above the ground). The canopy is the part of a forest in which the branches of the trees spread out and block sunlight from reaching the forest floor. This leafy environment is full of life in a tropical rainforest and includes insects, birds, reptiles, mammals, and more.

CAPILLARY ACTION. Capillary action is the movement of water as it is pulled upwards through tubes (xylem) within a plant's roots, stems, and leaves. The water (containing minerals and dissolved nutrients) is driven against gravity by adhesion of the water molecules (they stick to the sides of the tubes), cohesion of those molecules (the water molecules sticking together), and surface tension (the forces of the molecules on surface of the upward-moving water).

CAPSULE .A capsule is a seed pod that opens when it is dry and the seeds are mature.

CARBON DIOXIDE. Carbon dioxide, CO_2, is a molecule that has one carbon atom and two oxygen atoms; it is a gas at standard temperature and pressure. Plants use carbon dioxide gas in the photosynthetic process.

CARNIVORE.

Carnivores are animals that eat meat. They usually have

sharp teeth and powerful jaws.

CARPEL. The carpel is the female reproductive organ of a flower - it makes the seeds. It consists of the stigma, style and ovary. There may be more than one carpel in a flower.

CASPARIAN STRIP The Casparian strip is waxy layer (a band of suberin, a water-proofing material) that is located in the walls of plant root cells. This barrier strip stops the transport of water and minerals into the main vascular system of the root.

CAUDEX. The caudex is an enlarged, woody base of the stem or trunk (located just below the gound) on some plants - it is used for water storage. Many desert plants have a caudex, an adaptation to dry conditions. Some palms, cycads, and succulents have a caudex.
CAUDICIFORM Caudiciform means having a caudex.

CEDAR. Cedar trees (genus Cedrus) are large evergreen coniferous trees (up to about 80 feet tall) that have dense clusters of needles held in wide, woody peg-like structures.

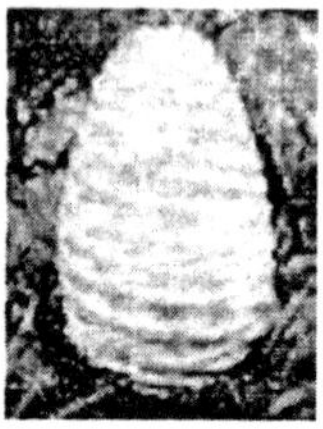

The barrel-shaped cones are held upright above a branch. Some cedars include: Cedar of Lebanon (*Cedrus libani*); Atlas cedar (*Cedrus atlantica*) – with blue-green foliage; Deodar cedar (*Cedrus deodara*) – with drooping branches.

CELL. The cell is the basic structural unit of all organisms. Plant cells have a tough outer cell wall, a cell membrane, genetic material (DNA), cytoplasm, and many organelles

CELL MEMBRANE. A cell membrane (also called the plasma membrane or plasmalemma) surrounds each plant cell (it is located within the cell wall)

CELLWALL. Plant cells have a thick, rigid cell wall located outside the cell membrane.

The cell wall is made of cellulose (a polysaccharide carbohydrate), proteins, and sometimes lignin. The cell wall gives the cell most of its support and structure. The cell wall also bonds with other cell walls to form the structure of the plant.

CELLULAR RESPIRATION. Cellular respiration is a process in which energy is produced from various molecules (like glucose), producing ATP (adenosine triphosphate). During cellular respiration, oxygen is used and carbon dioxide is produced Cellular respiration occurs in the mitochondria of eukaryotes, and in the cytoplasm of prokaryotes.

CELLULOSE Cellulose is a carbohydrate that comprises much of a plant's cell, especially the cell wall.

CENOZOIC ERA.

The "Age of Mammals" (65 million years ago to today), saw the emergence of familiar life forms, humans, the modern look of the continents, and a cooling climate. The Cenozoic (meaning "recent life") followed the Mesozoic Era.

CENTROSOME. A centrosome (also called the "microtubule organizing center") is a small body located near the nucleus - it has a dense center and radiating tubules. The centrosome is where microtubules are made. During cell division (mitosis), the centrosome divides and the two parts move to opposite sides of the dividing cell. Unlike centrosomes in animal cells, plant cell centrosomes do not have centrioles.

CEREAL. A cereal is a grain that is used for human food. Some cereals include rice, oats, wheat, and barley.

CERRADO .The cerrado is a grassy, treeless plain that surrounds the Brazilian rainforest.

cf. .Cf. is an abbreviation for "compare" or "compare with."

CHALK. Chalk is a soft, white type of limestone (a sedimentary rock). It consists mostly of calcium carbonate ($CaCO_3$) from ancient, microscopic, single-celled marine invertebrate shells. This type of rock is very porous, soft (a hardness of 3 on the Mohs scale), and crumbly. The chalk used to draw with is actually gypsum (calcium sulfate, $CaSO_4$-$2H_2O$)

CHARACTER. A character is a inherited trait of an organism. Characters are usually described in terms of a state, for example: "blue eyes" vs. "brown eyes," where "eyes" is the character, and "blue" and "brown" are its states.

CHICXULUB CRATER. The Chicxulub crater at the tip of the Yucatán Peninsula is an impact crater that dates from 65 million years ago. It is 120 miles wide and 1 mile deep. It is probably the site of the K-T meteorite impact that caused the extinction of the dinosaurs and many other groups of organisms.

CHLOROPHYLL. Chlorophyll is a molecule that can use light energy from sunlight to turn water and carbon dioxide gas into sugar and oxygen (this process is called photosynthesis). Chlorophyll is magnesium based and is usually green.

CHLOROPLAST. Chloroplasts are small green structures in plants that contain chlorophyll. Leaves have many chloroplasts.

CHONDRITIC METEOR. Chondritic meteors are stony meteors with chondrules, tiny glass spheres. These meteors are unchanged since their formation, shortly after the formation of the sun. These meteors consist of elements also common in the Earth's core.

.**CHROMOSOME.** Chromosomes are microscopic, self-replicating structures found in the nucleus of cells. Chromosomes contain genetic material (coiled stands of DNA that contain many genes). The genome of an organism is made up of the set of chromosomes that contain all of its genes. Chromosomes were discovered by Walther Fleming in 1879; the term chromosome was proposed by Waldeyer in 1888.

CINCHONA TREE. The cinchona tree is a tropical tree that is the primary source of the anit-malarial drug quinine. Quinine is found in the bark of the cinchona tree. Quinine is a chemical that cures malaria, a deadly tropical disease carried by mosquitoes. There are many species of cinchona; they range from about 15 to 20 meters tall. The cinchona tree is native to rainforests of the eastern slope of the Amazonian Andes of South America, where it is called the "fever tree." Classification: Family Rubiaceae, Genus Cinchona, Species C. officinalis, C. ledgeriana, C. uccirubra, C. calisaya, and others.

CLADODE. A cladode is a stem that looks like a leaf.

CLADE. A clade is a group of all the organisms that share a particular common ancestor (and therefore have similar features). The members of a clade are related to each other. A clade is monophyletic.

CLADISTICS. Cladistics is a method of classifying organisms based on common ancestry and the branching of the evolutionary family tree. Organisms that share common ancestors (and therefore have similar features) are grouped into taxonomic groups called clades. Cladistics can also be used to predict properties of yet-to-be discovered organisms.

CLADOGRAM. Cladograms are branching diagrams that depict species divergence from common ancestors. They show the distribution and origins of shared characteristics. Cladograms are testable hypotheses of phylogenetic relationships.

CLASS . In classification (taxonomy), a class is a group of related or similar organisms. A class contains one or more orders. A group of similar classes forms a phylum.

CLASSIFICATION . The classification of organisms helps in their study. Cladistics is a method based on common ancestry; the Linnean system is based on a simple hierarchical structure.

CLEFT. A cleft (also called parted) leaf is one in which the margins between the irregular teeth go more than halfway to the midrib.

CLOUD FOREST .A cloud forest is a rainforest that is on a mountainside. It is usually misty and cloudy.

CLIMAX VEGETATION .Climax vegetation is the final stage in the development of an area.

CLUB MOSSES

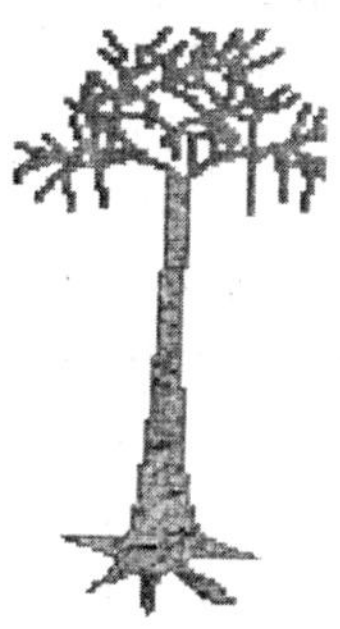

Club mosses (Lycopsids) are primitive, vascular plants (pteridophytes) that evolved over 375 million years ago (during the Devonian). Huge club mosses went extinct during the Permian mass extinction; smaller ones lived during the time of the dinosaurs. These plants live near moisture (in order for their spores to germinate). These fast-growing, resilient plants propagate with rhizomes (underground stems). Classification:

Division Lycopodiophyta, Class Lycopodiopsida, Order Lycopodiales, Family Lycopodiaceae (Club-mosses).

COAL. Coal is a combustible mineral formed from organic matter (mostly plant material) that lived about 300 million years ago (during the Pennsylvanian Period). During the Pennsylvanian Period, the earth was covered with huge swampy forests of giant ferns, horsetails, and club mosses. As layer upon layer of these plants died, they were compressed and covered with soil, stopping the decomposition process, forming peat. Heat and pressure chemically forced out oxygen and hydrogen, leaving carbon-rich deposits, called coal. A 20-foot-thick layer of plant material produces a one-foot-thick layer (seam) of coal

COLEOPTILE. A coleoptile is a protective sheath that surrounds the shoot tip and the embryonic leaves of the young shoot of grasses.

COMET. A comet is a celestial body that orbits around the sun. Its tail of gas and dust always points away from the sun.

COMMENSALISM. Commensalism is a situation in which two organisms are associated in a relationship in which one benefits from the relationship and the other is not affected much. The two animals are called commensals. An example pf commensalism is bromeliads (plants living on trees in rainforests) and frogs; the frogs get shelter and water from the bromeliad but the bromeliad is unaffected. Commensalism is a type of symbiosis.

COMPANION CELL. A companion cell is a type of cell that pumps nutrients (sugars) into phloem cells

COMPLETE FLOWER. A complete flower has a stamen, a pistil, petals, and sepals.

COMPOSITE FLOWER. A composite flower (like the sunflower) has a many individual flowers (called florets) on a wide, flat receptacle, that look like a single flower. The flowers in the central disk are called disk flowers; the flowers on the periphery are called ray flowers. This group is called Asteraceae (Compositae).

COMPOUND LEAF

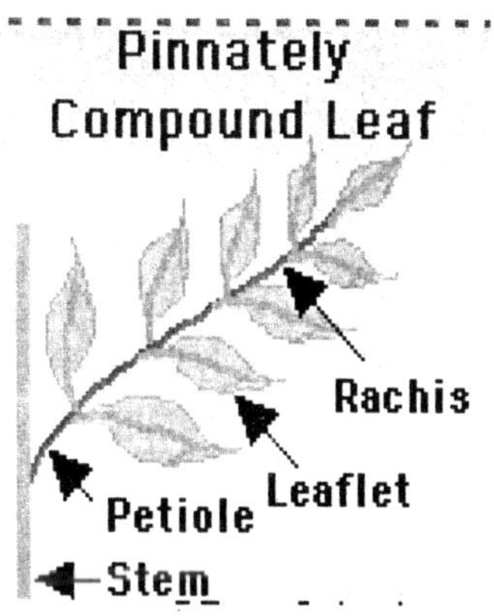

A compound leaf is a leaf that is divided into many separate parts (leaflets) along a midrib (the rachis). All the leaflets of a compound leaf are oriented in the same plane. When a compound leaf falls from the tree, it falls as a unit. A double compound leaf is one in which each leaflet of a compound leaf is also made up of secondary leaflets.

CONE.

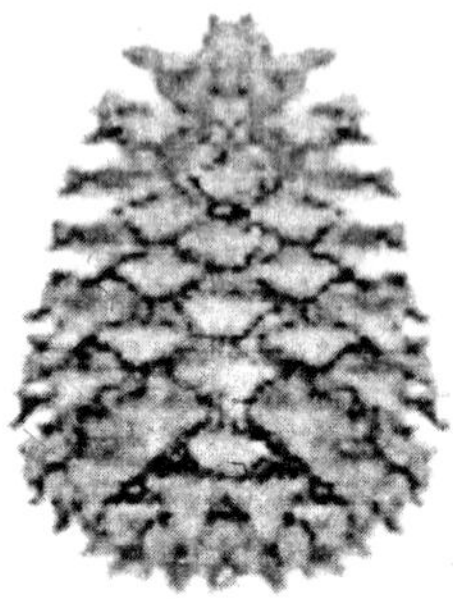

A cone (strobilus) is the reproductive fruiting structure of many tracheophytes. It is a group of scales that are joined to a central stalk; the seeds are borne on the surface of the cone scales. A cone scale contains either ovules or spores (depending on whether it is female or male).

CONIFER.

Most conifers are evergreen trees and shrubs that bear naked seeds in cones (a woody strobilus). Examples of modern-day conifers include pine, fir, larch, redwood, and spruce trees. Mesozoic Era conifers included redwoods, yews, pines, the monkey puzzle tree (Araucaria), cypress, and Pseudofrenelopsis (a Cheirolepidiacean). Towards the end of the Mesozoic, flowering plants flourished and began to overtake conifers as the dominant flora.

CONSERVATION. Conservation is the wise use of natural resources (plants, animals, minerals, water, etc.) so that they are not damaged and will be in good condition in the future.

CONSUMER. A consumer is a living thing that eats other living things to survive. It cannot make its own food (unlike most plants, which are producers). Primary consumers eat producers, secondary consumers eat primary consumers, and so on. There are always many more primary consumers than secondary consumers, etc.

CONTINENTAL DRIFT . Continental drift is the movement of the Earth's continents. The land masses are hunks of Earth's crust that float on the molten core. The ideas of continental drift and the supercontinent of Pangaea were presented by A. Wegener in 1915

CONTRACTILE ROOT. A contractile root is a root that contracts (gets smaller) and pulls down the crown of the plant below the surface of the soil.

CONVERGENT EVOLUTION. Convergent evolution (convergence) is when a trait develops independently in two or more groups of organisms. An example of convergence is the wings of Pterodactyls and bats.

COOKSONIA. Cooksonia is the oldest-known land plant. This primitive plant dates from Silurian period, about 428 million years ago Cooksonia was an erect plant with dichotomous branches and terminal sporangia (sacs that produce reproductive spores). Cooksonia fosils have been found in the USA, Canada, and Czechoslovakia

COPPICE SHOOT. A coppice shoot (also called a epicormic shoot, sap shoot, water shoot, or water sprout) is a shoot (new growth) that arises from an adventitious or dormant bud on a branch or a stem of a plant (usually near the base of the plant). This fast-growing shoot often starts to grow when part of a forest canopy is removed or thinned (allowing light in).

COPPICE STAND. A coppice stand is an area of coppice shoots.

CORN.

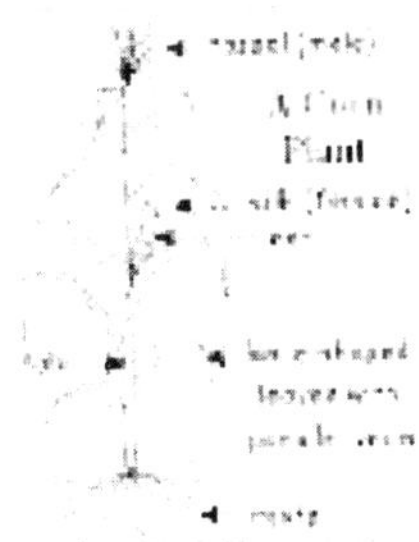

Corn is a cereal plant in the grass family. It an important agricultural plant native to Mexico or South America. It's scientific name is *Zea mays*. The edible part of corn in the kernels, rows of seeds that

grow on an elongated ear covered by husks (specialized leaves).

CORPSE FLOWER. The "corpse flower" is the world's largest flower. This giant bloom is found in rainforests of Indonesia. It's scientific name is *Rafflesia arnoldi*. Rafflesia gives off a putrid smell that reminds people of rotting meat (this odor attracts its pollenators, beetles and flies), hence its nickname. Rafflesia's enormous flower is about 3 feet (1 m) across and weighs about 20 pounds (9 kilograms). The flower takes about a year to develop, then it blooms for about a week before dying. The flower has five wide orange petals (with pale dots) surrounding a spiked cup. Rafflesia has no stem, no roots, and no leaves. The flower is supported by fungus-like tissue that lives in another plant - the Tetrastigma vine.

COQUINA. Coquina is a type of limestone (a kind of sedimentary rock) that is mostly made of shells and shell fragments.

CORDATE LEAF.

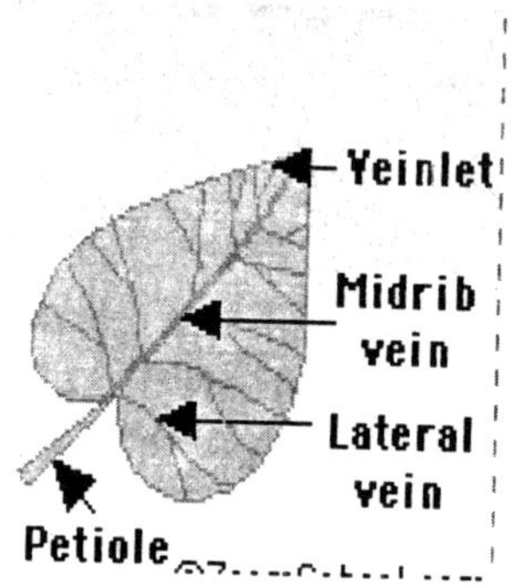

A cordate leaf has a heart shape, with the wide part towards the petiole.

CORK. Cork (also called periderm) is the soft, light-weight bark of the cork oak tree. This low-density material floats in water. Cork cells are made by cork cambium cells. Cork contains suberin, a waxy, water-proof material. Cork protects the tree from water loss and from insects and infections.

CORM. A corm is a fleshy underground stem of some plants. It looks like a bulb, but is solid (it is not formed in layers).

CORN. Corn (*Zea mays*), also called maize, is a type of cereal grass; it is an edible grain. This tall, annual plant has long, alternately-spaced blade-like leaves, and a strong, solid stem. A flowering plant, staminate (male) flowers grow on the tassels at the end of the main axis of the stem. The pistillate (female) inflorescence grows into the ear of corn, and is a spike having a thick axis paired spikelets in rows (each row of paired spikelets produces two rows of grain). The ear is covered by modified leaves, called husks or shucks. Corn evolved in the Americas, but has been brought all around the world by people; it is the second-largest food crop (behind wheat). Classification: Division Magnoliophyta (angioperms), Class Magnoliopsida (dicots), Class Liliopsida (monocots), Subclass

Commelinidae (grasses, sedges and rushes), Order Cyperales, Family Poaceae (Gramineae) (grasses).

COROLLA. The corolla consists of the petals of a flower.

COTYLEDON. The cotyledon is the embryonic leaf within a seed. When a seed germinates, the cotyledon is the first leaf to grow. Monocots have one cotyledon; dicots have two cotyledons

CRENATE.

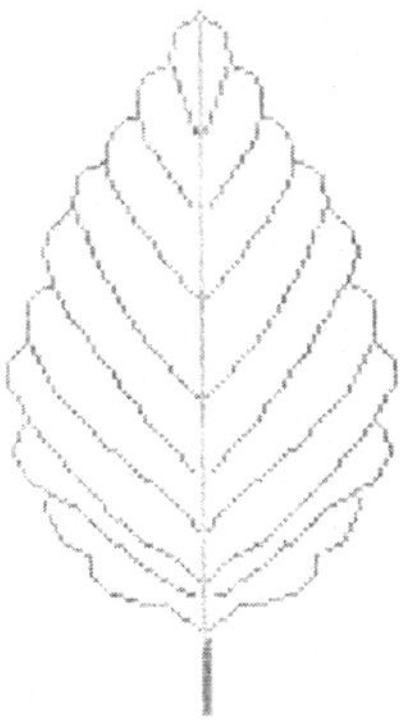

crenate

A crenate leaf has edges (margins) shaped like rounded teeth.

CRENOLATE. A crenolate leaf margin has edges that are shallow-toothed.

CRISTAE. singular crista) The multiply-folded inner membrane of a cell's mitochondrion that are finger-like projections. The walls of the cristae are the site of the cell's energy production (it is where ATP is generated).

CROSS-POLLINATION. Cross-pollination is the transfer of pollen from the anther to the stigma of a flower on a different plant.

CYTOPLASM. Cytoplasm is the jelly-like material outside the cell nucleus in which the organelles are located (the entire contents of the cell located inside the plasma membrane, but excluding the nucleus).

CYTOSOL. Cytoplasm is the jelly-like material in cells, excluding the cell nucleus and all organelles (the entire contents of the cell located inside the plasma membrane, but excluding the nucleus and other organelles).

CRETACEOUS PERIOD.

Dinosaurs were at their height during the Cretaceous period, 146-65 million years ago, and flowering plants spread and flourished. There was a mass extinction (the K-T mass extinction) at the end of the Cretaceous, marking the end of the dinosaurs and many other species of animals and plants.

CRISPED. Crisped leaves have a tighly curled margin. Parsley and kale leaves are crisped.

CROZIER .The crozier is the spirally coiled "fiddlehead" of a young fern leaf.

CROWN. The crown of a plant is the area from which new shoots arise or the point at which the roots meet the stem. Also, the upper area of the tree that has a lot of branching and leaves.

CRYPTOGAM. Cryptogams are plants and plant-like organisms that do not reproduce with seeds and do not produce flowers. Many cryptogams reproduce using spores. Ferns, mosses, fungi, and algae are cryptogams.

CRYSTALS. Crystals are solids whose atoms form a very regular pattern.

CULM. A culm is the elongated straw or hollow stem of grasses. The culm usually supportes the inflorescence.

CUNEATE. Cuneate means wedge-shaped.

CULTIVAR. A cultivar is a plant that is a **culti**vated (bred) **var**iety.

CUTICLE. The cuticle is the fatty or waxy outer layer of epidermal cells that are above ground.

CYANOBACTERIA. Cyanobacteria Blue-green algae (also called blue-green algae) are simple, (usually) one-celled photosynthetic organisms that lack a membrane-bound nucleus (they are prokaryotic). They belong to the kingdom Monera.

CYCAD

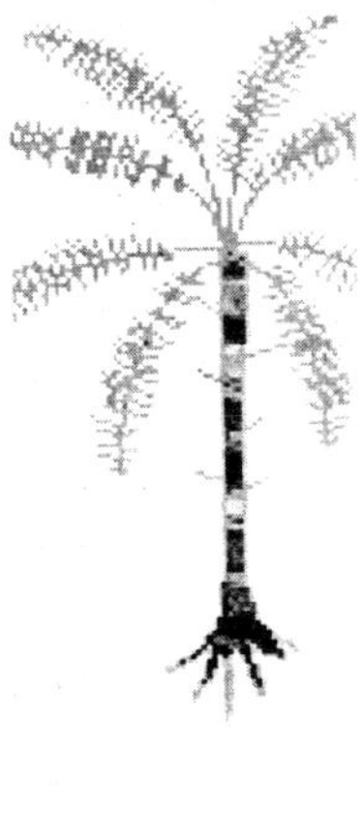

Cycads (Cycadophyta) are primitive seed plants that dominated the Jurassic period (cycads comprised 20% of the world flora). Cycads are palm-like trees that live in warm climates. Separate male and female plants exist (they are dioecious). These gymnosperms have long, divided leaves and produce large cones. Cycads evolved during the Pennsylvanian, had their heyday during the Mesozoic, and only about 185 species (in 11 genera) still exist today. Leptocycas (shown above) and Ptilophyllum were Mesozoic Era cycads. Later cycads had a more rounded, barrel-like base. Classifcation: Division Pinophyta (Gymnosperms), Subdivision Cycadicae, Class Cycadopsida, Order Cycadales, Family Cycadaceae (Cycads)

CYCADEOID. Cycadeoids (Bennettitales) were plants with woody stems (some erect, some spherical) and very tough leaves. Cycadeoids do not always have separate male and female plants. Cycadeoids are now extinct. Some Mesozoic Cycadeoids included:

Cycadeoidea, Vardekloeftia, Williamsonia (shown above), Williamsoniella, Westersheimia, and Leguminanthus

CYCADOPHYTES.

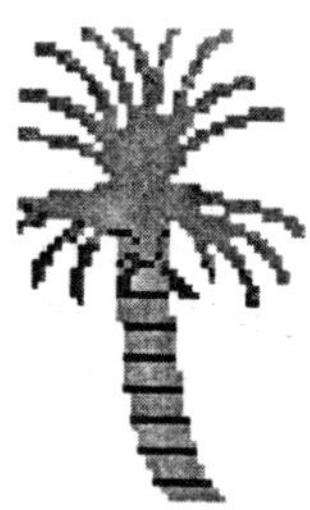

Cycadophytes included the Cycads and Cycadeoids (Bennettitales), plants with woody stems (some erect, some spherical) and very tough leaves. These two groups differ mainly in the way they reproduce: Cycads have separate male and female plants; Cycadeoids do not always. Cycadeoids are now extinct but there are still a few cycads. Some Mesozoic Era Cycads included: Leptocycas, Cycas, Zamia, Dioon, Bowenia, Stangeria, and Microcyas. Some Mesozoic Cycadeoids included: Cycadeoidea, Vardekloeftia, Williamsonia, Williamsoniella, Westersheimia, and Leguminanthus.

CYME. A cyme is an inflorescence where the central flower opens first.

CYTOPLASM.

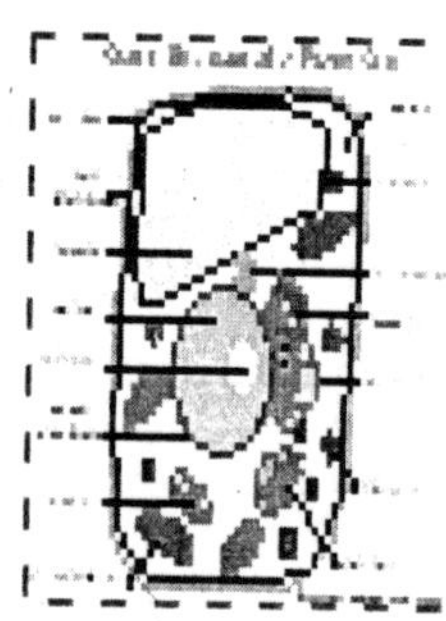

Cytoplasm is a clear, jelly-like fluid found inside cells.

CYTOSKELETON. The cytoskeleton is a network of protein filaments and microtubules that are located in the cytoplasm of eukaryotic cell, just under the cell membrane. The cytoskeleton acts as a support for the cell, as a spatial organizer, and as a channel for some chemical transport.

D

DATING FOSSILS. Dating a fossil is determining when that organism was alive. Paleontologists use many ways of dating individual fossils in geologic time, including stratigraphy, observations of the fluctuations of the Earth's magnetic field, radioisotope-dating, and looking at nearby index fossils.

DECIDUOUS. Deciduous plants lose their leaves seasonally, usually for the dry season. Some deciduous plants include ash, beech, hickory, maple, and oak.

DECOMPOSER. Decomposers are organisms like fungi and some bacteria that break down and digest the remains of organisms.

DECOMPOSITION. Decomposition is the decay or breakdown of things into more basic elements. For example, after a plant dies, it decomposes into organic nutrients.

DEFORESTATION. Deforestation is the loss of forest. Deforestation has many causes, including man's cutting down trees, forest fires, severe drought, changes in sea level, disease, etc.

DEHISCENT. A dehiscent is a structure on some plants that opens to release seeds or pollen grains

DEHISCENT FRUIT. A dehiscent fruit splits open when it is maturite, causing the dispersal of its seeds. Some dehiscent fruits include cotton, poppy, peanuts, milkweed, magnolia, and all beans

DEHYDRATE. To dehydrate is to lose a lot of water. Plants can become dehydrated in dry, hot weather

DENDROCHRONOLOGY. Dendrochronology is the science in which tree rings are studied to determine conditions in the past.

DENDROCLIMATOLOGY Dendroclimatology is the science in which tree rings are studied to determine climate changes in the past.

DESERT.

A desert is a very dry area that receives less than 10 inches (25 cm) of rainfall each year. Desert organisms have adapted to life with little water.

DETRIVORE. A detrivore (or detritus feeder) is an organism that eats the dead remains of other organisms (detritus). Examples of detrivores include some bacteria and fungi.

DIASPORE. A diaspore (also called a disseminule) is a part of a plant that is separated from the plant and dispersed (sent or taken away from the plant) for reproduction. Diaspora include seeds, fruits, and spores.

DICHOTOMOUS KEY. A dichotomous key is a method for determining the identity of sometihing (like a butterfly, a plant, or a rock) by going through a series of choices that lead the user to the correct name of the item. At each step of the process, the user is given two choices; each alternative leads to another questions until the identification is completed. For example, a question in a dichotomous key for trees might be something like, "Does it have flat or needle-like leaves?" Dichotomous means "divided in two parts".

DICHOTOMOUS VENATION. Dichotomous venation is a pattern of leaf veins in which the veins branch in two over and over again. Ferns are dichotomousy veined. A few angiosperms and gymnosperms (like gingkos) have dichotomous venation.

DICOT. A dicot (Class Magnoliopsida) is a type of flowering plant (an angiosperm) whose seed has two embryonic leaves (cotyledons). The leaf veins are usually net-like (and not parallel). Taproots are often present. Beans and peas are examples of dicots.

DICTYOSTELE. A dictyostele is a siphonostele (a cylinder of vascular tissue) that has two or more overlapping leaf gaps (as in ferns). It is a divided stele located in the outer cortex (instead of in the center of the stem). Dictyosteles are composed of individual vascular bundles

DIMORPHISM. Dimorphism means having two forms. It usually means that an organism has two different types, for example, males and females of some species look different, and are said to be sexually dimorphic.

DIOECIOUS. A dioecious plant has the male and female flowers on different plants. For example, date trees are dioecious. (Compare with monoecious.)

DIPLOID. A diploid cell has the same number of chromosomes that most other cells of that organism have (except the gametes, like the sperm and the egg, which are haploid).

DIRT.

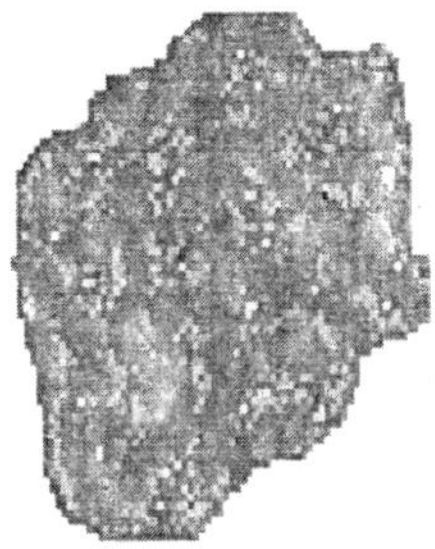

Dirt is another name for soil. Soil is a natural, constantly-changing substance that is made up of minerals, organic materials, and living organisms. Plants grow in soil.

DISPERSAL. Dispersal is the process in which an organism spreads out geographically. Seeds are dispersed by the winds and by animals.

DISPERSERS. Dispersers are animals that spread plant seeds. Some dispersers include birds, insects, bats, and furry animals like monkeys.

DISSEMINULE. A disseminule (also called a diaspore) is a part of a plant that is separated from the plant and dispersed (sent or taken away from the plant) for reproduction. Dissemules include seeds, fruits, and spores.

DNA.

DNA (short for deoxyribonucleic acid) is a complex organic molecule that carries the genetic information of an organism. DNA is the primary constituent of chromosomes. DNA is made up of two strands of amino acid bases (adenosine, cytosine, guanine, and thymine) arranged in a double helix. DNA's chemical instructions are carried to throughout the cell by another nucleotide, RNA.

DOGGER EPOCH. The Dogger epoch was the middle part of the Jurassic period, about 180 to 159 million years ago.

DOMINANT. The dominant plant is the most adundant species in an area, for example, pine trees are dominant in a pine forest.

DORMANCY. Dormancy is a period in which a plant has no active growth in response to harsh environmental conditions (like droughts or cold seasons).

DOUBLE COMPOUND LEAF. A double compound leaf is a leaf in which each leaflet of a compound leaf is made up of secondary leaflets.

DRIP TIP. The drip tip is the long, sharply-pointed tip on some rainforest plant leaves that allows water to flow quickly off the leaf surface. In the rainforest, many leaves have a drip tip. Without it, water would build up on the leaf.

E

EARTH.

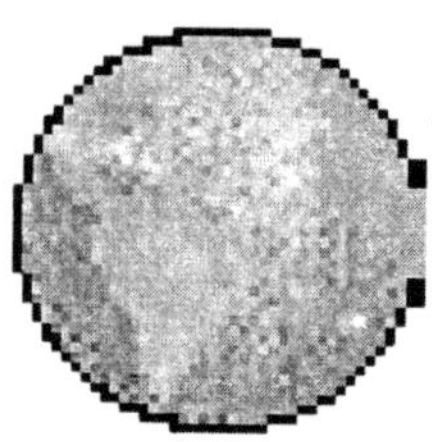

The Earth is the third planet from the Sun.

ECOSYSTEM. An ecosystem is the interrelationships between all of the living things in an area.

EDICARA FAUNA. Ediacara fauna is the animal life that lived during the Vendian or Ediacaran period (roughly 650 to 544 million years ago). The Vendian is when the earliest-known animals evolved. Vendian biota (Ediacara fauna), included soft-bodied multi-cellular animals, like sponges, cnidarians, worms, and soft-bodied relatives of the arthropods. The Ediacara was named for the Ediacara Hills in Australia, north of Adelaide, where these early animal fossils were first found (in 1946, by the Australian mining geologist Reginald C. Sprigg). Other Vendian Period fossils have been found in Mistaken Point, Newfoundland and the White Sea off the northern coast of Russia.

EMBRYO. An embryo is a developing plant still inside the seed. The embryo has cotyledons (embryonic leaves), a root cap, a food source and a plumule (shoot), all located inside the protective seed coat.

EMERGENTS . Emergents are giant trees in a rainforest that are much higher (up to 270 feet or 81 m) than the average canopy height. The emergents house many birds and insects in a very dry environment.

ENDANGERED SPECIES.

An endangered species is a group of organisms that is dwindling in numbers and may go extinct soon. Many species of plants are endangered.

ENDEMIC. Endemic plants are native to an area and are only found in that area.

ENDOPLASMIC RETICULUM. The endoplasmic reticulum (ER) is a network of membranes in plant cells (and all eukaryotic cells) that control protein synthesis and cellular organization.

ENTIRE. An entire leaf has smooth edges (margins) (with neither teeth nor lobes).

entire

EPHEMERALS. Ephemerals are plants that have an accelerated life cycle. These hardy plants grow and reproduce quickly - they are often dormant during extreme weather (droughts, heat spells, cold spells, etc.). Their seeds are drought resistant. These plants often live in areas with harsh climates, like deserts and very cold areas. Some ephemerals include: Desert Paintbrush, Desert Sand Verbena, Dutchman's Breeches, Mojave Aster, and Yellow Trout Lily.

Phanerozoic Eon. 540 million years ago through today

Proterozoic Eon. 2.5 billion years ago to 540 million years ago Archaeozoic Eon 3.9 to 2.5 billion years ago

Hadean Eon. 4.6 to 3.9 billion years ago

EPICORMIC SHOOT. An epicormic shoot (also called a coppice shoot, sap shoot, water shoot, or water sprout) is a shoot (new growth) that arises from an adventitious or dormant bud on a branch or a stem of a plant (usually near the base of the plant). This fast-growing shoot often starts to grow when part of a forest canopy is removed or thinned (allowing light in).

EPICOTYL. The epicotyl is the part of the stem that is above the first leaves.

EPIDERMIS. The epidermis is the outer protective layer of a plant. This tissue helps prevent injury and minimizes water loss by evaporation.

EPIGYNOUS OVARY. An epigynous ovary (also called an inferior ovary) is an ovary located below the flower parts (the calyx, corolla, and androecium). The flower parts are attached to the top of the ovary.

EPIPHYTE. Epiphytes are plants that live attached to a plant (or other structure like a rock, telephone pole or a building) and not in the ground). Epiphytes include many orchids and bromeliads. Epiphytes are not parasites; they get water and nutrients from the air (and not their host).

EPOCH. An epoch is a division of a geologic period; it is the smallest division of geologic time, lasting several million years.

EQUATORIAL RAINFOREST. Tropical rainforests are warm, very wet forests that do not freeze (the temperature remains over 75°F=24°C) and do not get extremely hot. Tropical rainforests cover about 7% of the Earth's surface, in a band around the equator. They are also called tropical rainforests.

EQUISETUM. Equisetum (a modern genus of horsetail) is a primitive, spore-bearing plant (a sphenopsid) with rhizomes. Its side branches are arranged in rings along the hollow stem. Other genera of horsetail were common during the Mesozoic Era, like Neocalamites, Calamites etc. Horsetails date from the Devonian period 408-360 million years ago, but are still around today and are invasive weeds.

ER. ER stands for endoplasmic reticulum. The ER is a network of membranes in plant cells (and all eukaryotic cells) that control protein synthesis and cellular organization.

ERA. Two or more geological periods comprise an Era, which is hundreds of millions of years in duration.

ETHNOBOTANY. Ethnobotany is the science which studies how plants are used in various cultures.

EUKARYOTE. A eukaryote (which means "true nucleus") is an organism whose cells have internal membranes (which divide the cell into regions that have different functions) and a structurally-discrete nucleus. They also have a cytoskeleton which controls their shape. Eukaryotes include plants, animals, fungi, and protists (they do not include blue-green algae, bacteria, viruses, and other primitive microorganisms).

EURASIA. Eurasia is the combined, joined land masses of Europe and Asia.

EUTROPHICATION. Eutrophication is when the concentration of nutrients in a waterway increases; this occurs when sewage, fertilizers, or sediments enter the water. This increase in the concentration of nitrogen and/or phosphorous may result in an algal bloom (an overabundance of blue-green algae that depletes the water of oxygen, killing other organisms).

EVAPORATION. Evaporation is the process in which a liquid is transferred into gaseous form.

EVAPOTRANSPIRATION. Evapotranspiration is the process in which rain water evaporates from trees and returns to the atmosphere.

EVERGLADES. The Everglades is a swamp located in southwestern Florida. It contains over 2,000 different types of plants, including saw grass (Cladium jamaicense), mangroves (including the red, black and white mangroves), alligator flag (Thalia geniculata), strangler fig (Ficus aurea), gumbo-limbo (Bursera simaruba), mahogany (Swietenia mahagoni), saw palmetto, pond cypress (Taxodium ascendens), bald cypress (Taxodium distichum), moonvine (a morning glory), coontie, and various willows, slash pine (Pinus elliottii densa) and other pines, and oaks with Spanish moss hanging form the limbs. Many epiphytes (air plants) like the night-blooming epidendrum (Epidendrum nocturnum) also live in trees.

EVERGREEN.

Evergreen plants do not lose their leaves seasonally. Pines and firs are examples of evergreens. (Compare with deciduous plants.)

EVOLUTION. Evolution is a process in which the gene pool of a population gradually (over millions of years) changes in response to environmental pressures, natural selection, and genetic mutations. All forms of life came into being by this process.

EVOLVE. To evolve is to develop by the process of evolution, changing in some way as an adaptation to the environment.

EXTINCT. An animal species that is extinct has died out. Most animal species that ever existed have gone extinct, including all the dinosaurs.

EXTINCTION. Extinction is the process in which groups of organisms (species) die out.

EXTINCTION-LEVEL EVENT. An Extinction-Level Event is a catastrophic event (such as a large asteroid or comet hitting the Earth, a large change in the Earth's temperature/sea level, tremendously increased volcanism, etc.) that is capable of causing a mass extinction. This event would greatly damage the ecosphere of Earth, causing many groups of organisms to die.

EXTRACTIVE RESERVE. An extractive reserve is an area in a rain forest in which people are allowed to harvest crops, such as rubber or Brazil nuts.

EYE. An eye is an axillary bud on an underground plant part such as a tuber (e.g., potato) or in the area where the stem joins the tuberous root (e.g., dahlia).

F

FAMILY. In classification, a family is a group of related or similar organisms. A family contains one or more genera (plural of genus). A group of similar families forms an order.

FAST PLANTS. Fast Plants are fast-growing plant varieties that were developed at the University of Wisconsin, Madison, Wisconsin, USA. Fast Plants complete their life cycles about 45 days after planting. Fast plants are used in classrooms and laboratories around the world for teaching and research. The first plants in space were Fast Plants.

FEN. A fen is a wet area rich in peat and other organic matter.

FERN. Ferns are non-flowering vascular plants with rhizomes that were plentiful during the Mesozoic Era and usually live in warm, moist areas. Ferns have fronds divided into leaflets. Classification: Plyla: Lycophyta (lower ferns like clubmosses), Pterophyta (ferns), Psilophyta (whisk ferns) (true ferns belong to the Class Filices).

FERTILIZER. Fertilizer is material that is added to soil to increase is fertility and output. Fertilizers include manure, compost, and chemical mixtures.

FIBROUS ROOT. A fibrous root is a type of primary root of a plant that has a lot of side branching (compare with tap root). Fibrous root are the most common type of roots.

FILAMENT. The filament is the part of the flower that holds the anther.

FLAGELLUM. A flagellum is a long, thin, whip-like structure that is found on some organisms, including some unicellular algae. Flagella help propel the organism through water.

FLAX. Flax (family Linaceae, genus Linum) is a flowering plant from cool, temperate areas. It has narrow, lance-shaped leaves and blue flowers - the plant grow to be up to about 1.5 to 2 feet tall. Flax is grown for its seeds (which yield oil) and linen fiber (a strong vegetable fiber that is made from the woody stalk of the flax plant).

FLORA. The flora is all the plants that live in an area.

FLOWER.

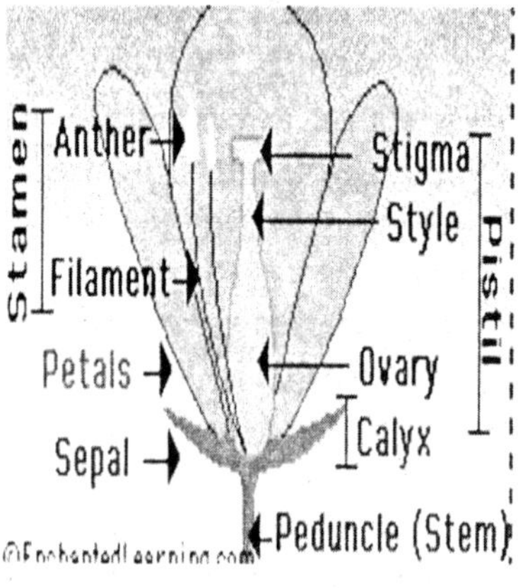

The flower is the reproductive unit of angiosperms. Flowers usually have carpels, petals, sepals, and stamens. Some flowers (called perfect flowers) have both male and female reproductive organs; some flowers (called imperfect flowers) have only male reproductive organs (stamens) or only female reproductive organs (ovary, style, and stigma). Some plants have both male and female flowers, while other have males on one plant and females on another. Complete flowers have a stamen, a pistil, petals, and sepals. Incomplete flowers lack one of these parts.

FLOWERING PLANTS.

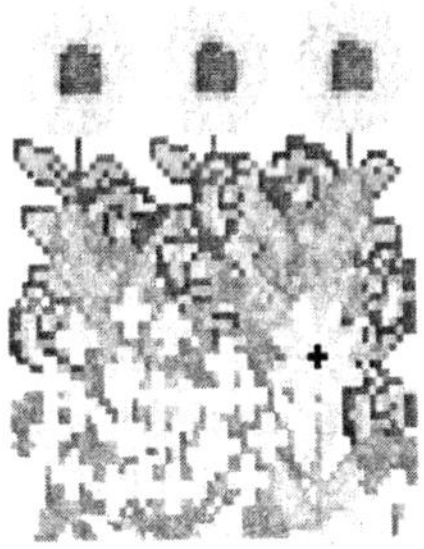

Flowering plants are Angiosperms (meaning "covered seed"). They produce seeds enclosed in fruit (an ovary). They are the dominant type of plant today; there are over 250,000 species, including grasses, peas, etc. Their flowers are used in reproduction. Angiosperms evolved about 140 million years ago, during the late Jurassic period, and were eaten by dinosaurs. They became the dominant land plants about 100 million years ago (edging out conifers, a type of gymnosperm). Angiosperms are divided into the monocots (like corn) and dicots (like beans).

FOLIAGE.

Foliage is the leaves of plants.

FOLIVORE. A folivore is animal that eats foliage (leaves). For example, the green iguana is a folivore.

FOOD WEB.

A food chain is is the sequence of who eats whom in a biological community (an ecosystem) to obtain nutrition. A food web is all of the interactions between predators and prey in which plants and animals obtain food in an ecoystem. The web starts with plants or other autotrophs (organisms that make their own food from light and/or chemical energy) that are eaten by herbivores (plant-eaters). The herbivores are eaten by carnivores (meat-eaters). These are eaten by other carnivores. When any organism dies, it is eaten by tiny microbes (detrivores) and the exchange of energy continues.

FORB. A forb is a small, herbaceous (non-woody), broad-leaved vascular plant (excluding grasses, rushes, sedges, etc.). For example, wild flowers are a type of forb.

FOREST. A forest is an area in which trees are the dominant plant

FOREST FLOOR. The forest floor is the lowest layer of a rainforest, extending from the ground to about 3 feet (1 m) high. This layer is teeming with animal life, especially insects. The largest animals in the rainforest generally live here

FOREST SUCCESSION. Forest or plant succession is the natural pattern of ecosystem growth and change over time for a particular environment. Plant life follows established patterns of growth and change after major distruptions, like fires, floods, agricultural damage, logging, etc. Generally, smaller, fast-growing herbaceous species and grasses grow first in an open field, followed in a few years by softwood tree seedlings and larger herbaceous species. As a young forest develops into a mature forest (30 to 70 years), an understory of smaller hardwood trees develops. The final stage is a climax hardwood forest (100 plus years).

FOSSIL. Fossils are mineralized impressions or casts of ancient animals and plants. Fossils have been found on every continent on Earth.

FOSSIL FUEL. A fossil fuel is a naturally-ocurring, energy-rich organic (carbon-based) substance (like shale, petroleum (oil), coal, or natural gas) in the Earth's crust that was formed from ancient organic material (mostly plants).

FROND . A frond is the fern-like (or feathery) foliage of a plant that has many divisions. Ferns and palms have fronds. Frond is also used to refer to the main part of a kelp plant (excluding the holdfast).

FRUIT. A fruit is the part of a flowering plant that contains the seeds. Some fruits include apples, oranges, berries, maple pods, and acorns. Some fruit are fleshy and some are dry, like cotton (a dehiscent fruit) and sunflower (an indehiscent fruit). Not all fruit are edible. True (simple) fruits (like the tomato, coconut, watermelon, olive, lemon, and banana) develop from the wall of a single ovary. False (compound) fruits (like the strawberry, rosehip, and pineapple) develop from more than one pistil.

FUNGUS.

Fungus (plural fungi) are organisms that obtain energy by breaking down dead organic material and that produce spores. Some fungi include mushrooms, toadstools, slime molds, yeast, penicillium mold, and mildew. Classification: kingdom Fungus.

G

GAMETANGIUM. (plural gametangia) A gametangium is a reproductive organ that is in some plants (especially algae, fungi, mosses, and ferns). The gametangium produces gametes (reproductive cells).

GAMETE. A gamete is the male or female reproductive cell of an organism (the sperm or the egg). Each gamete has only half the number of chromosomes that the other cells of that organism have (it is haploid).

GALLIC EPOCH. The Gallic epoch was the middle part of the Cretaceous period, about 127 million to 89 million years ago.

GASTROLITHS. (pronounced GAS-trohliths) Gastroliths are stones that some animals swallow and use to help grind up tough plant matter in their digestive system. They're also called gizzard rocks.

GENOME. The genome of an organism is made up of the set of chromosomes that contain all of its genes.

GENUS. (pronounced GEE-nus) In classification, a genus is a group of related or similar organisms. A genus contains one or more species. A group of similar genera (the plural of genus) forms a family. In the scientific name of an organism, the first name is its genus (for example, people are *Homo sapiens* - our genus is *Homo* and our species is *H. Sapiens*).

GEOLOGICAL TIME. The history of the earth is described in geological time, which is measured in millions of years and billions of years. The divisions used are: eon, era, period, and epoch.

GEOLOGICAL TIME PERIODS. Geologic time is divided into divisions based on some distinguishing feature of that time (like an Ice Age). The divisions used are: eon, era, period, epoch, and age.

GEOLOGY. Geology is the study of the Earth's structure, including rocks.

GEOTROPISM. Geotropism is a plant's reaction to gravity in which the roots go towards the pull of gravity, and the shoots go in the opposite direction.

GERMINATION. Germination is the beginning of growth of a plant from its seed.

GIGANTOPTERID. Gigantopterids were ancient seed plants that lived during the Permian period, over 250 million years ago. It is thought that the gigantopterids evolved into the flowering plants. One species of gigantopterid is the broad-leafed climbing plant *Vasovinea tianiia*; it was found in 1993 in China.

GINKGO or GINGKO. Ginkgo or Gingko (also called the maidenhair tree) is a primitive seed-bearing tree (a gymnosperm) that was common during the Mesozoic Era, but has only one existing species now. Ginkgos peaked during the Jurassic and Cretaceous periods. This deciduous (losing its leaves in cold weather) tree has fan-shaped leaves divided into two lobes. Classification: Division Pinophyta (Gymnosperms) , Subdivision Pinicae, Class Pinopsida, Order Ginkgoales, Family Ginkgoaceae (ginkgos).

GIZZARD ROCKS. Gizzard rocks are stones that some animals swallow and use to help grind up tough plant matter in their digestive system. They're also called gastroliths.

GLABROUS. A glabrous surface lacks hairs (and has a smooth surface).

GLOCHID. Glochids are tufts of short, barbed spines that are found at the areoles of opuntia cacti.

GLOSSOPTERIS.

Glossopteris (from the Greek glossa, meaning tongue, because the leaves were tongue shaped) is a genus of extinct seed fern (a Pteriosperm) whose fossils are found throughout India, South America, southern Africa, Australia, and Antarctica. Glossopteris was about 12 ft (3.6 m) tall. The distribution of this fossil plant throughout the southern hemisphere led the Austrian geologist Eduard Suess to deduce that there had once been a land bridge between these areas. He named this large land mass Gondwanaland (named after a district in India where the plant Glossopteris was found). This was the southern supercontinent formed after Pangaea broke up during the Jurassic period. It included what are now the continents South America, Africa, India, Australia, and Antarctica. These deciduous (losing their leaves in the cool season) gymnosperms arose during the late Permian and became dominant, but went extinct by the end of the Triassic period.

GOLGI BODY. (also called the Golgi apparatus or Golgi complex) a flattened, layered, sac-like organelle that looks like a stack of pancakes and is located near the nucleus. It produces the membranes that surround the lysosomes. The Golgi body packages proteins and carbohydrates into membrane-bound vesicles for "export" from the cell.

GONDWANALAND. Gondwanaland, also known as Gondwana, was the southern supercontinent formed after Pangaea broke up during the Jurassic period. It included what are now the continents South America, Africa, India, Australia, and Antarctica. Gondwanaland was named for a district in India where the fossil plant Glossopteris was found; this plant led E. Suess to deduce that the southern continents were once joined, supporting Wegener's continental drift theory.

GRAFT. A graft is a shoot or bud that has been joined to another plant.

GRAIN. A grain is a single particle of pollen.

GRANUM. (plural grana) A stack of thylakoid disks within the chloroplast is called a granum.

GRASS. Grasses are a group of flowering plants (angiosperms) that belong to the family Graminae. Classification: Division Magnoliophyta (angioperms), Class Magnoliopsida (dicots), Class

Liliopsida (monocots), Subclass Commelinidae (grasses, sedges and rushes, Order Cyperales, Family Poaceae (Gramineae) (grasses).

GRAZER. A grazer is an animal that eats low-lying vegetation, such as grasses and other low plants.

GRAZING FOOD CHAIN. The grazing food chain is a model that describes the flow of organic energy through organisms in an ecosystem. A trophic level is a level of this grazing food chain. Plants (called primary producers) occupy the first trophic level. Plant-eaters (also called primary consumers) occupy the second trophic level in the grazing food chain. The third level contains animals that eat primary consumers (first-level carnivores = secondary consumers). The fourth trophic level contains tertiary consumers, generally larger animals (like eagles) that eat primary consumers.

GREENHOUSE EFFECT. The greenhouse effect is an increase in the temperature of a planet as heat energy from sunlight is trapped in the atmosphere. Excess carbon dioxide and water vapor increase this effect. The greenhouse effect is strong on Earth, maintaining (and possible exacerbating) warm temperatures.

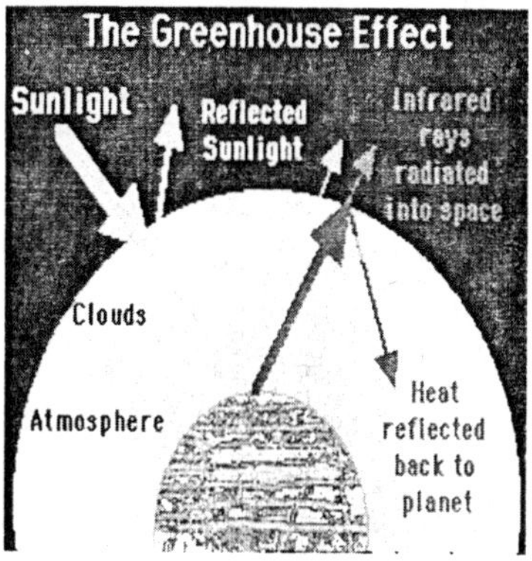

GREEN REVOLUTION. The green revolution was a dramatic increase in agricultural yields that occurred in the 1950s through 1960s. The green revolution was based upon many improvements in plant science, including the genetic improvement of many plants (including new, high-yield hybrid varieties), improved irrigation, more efficient machinery, new fertilizers, and pest controls that increased plants' disease-resistance, improved their hardiness, and increased their productivity (especially rice, wheat, and corn)

GUARD CELL.

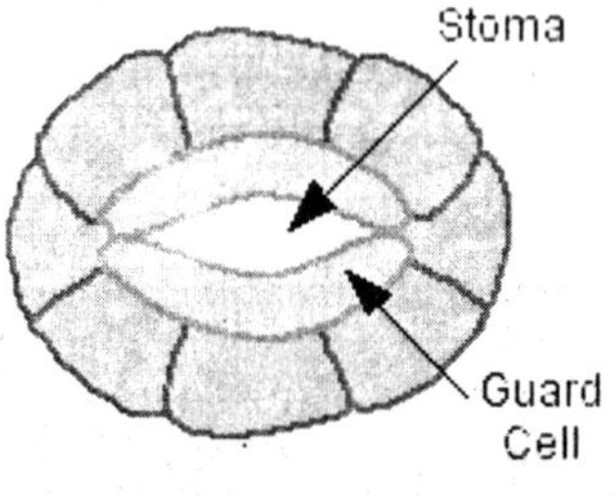

Each stoma has two crescent-shaped guard cells that control the size of the opening of the stoma using turgor pressure. This changes the amount of water vapor and other gases that can enter and leave the plant.

GUM. Gum is a sticky substance that is secreted by some plants. Gum hardens when it dries.

GYMNOSPERM.

Gymnosperms (meaning "naked seeds") are seed-bearing plants that don't produce flowers. These plants release pollen into the air to the female ovule, causing fertilization. Their seeds develop without a protective covering. The earliest gymnosperms were seed ferns from the Devonian period (408-360 million years ago). Some examples of gymnosperms are conifers (like pines, redwoods, and fir), gingkos, seed ferns, cycadeoids, and cycads. These plants were very important to plant-eating dinosaurs.

H

HABITAT. A habitat is a space (which includes food, water and shelter) suitable for the survival and reproduction of an organism.

HALF-LIFE. The half-life of a radioisotope is the amount of time it takes for half of the radioisotope to decay.

HALOMORPHIC. Halomorphic characteristics are adaptations to a saline (salt water) environments.

HALOPHTYE. A halophyte is a plant that has adapted to salty soils.

HAPLOID. A haploid cell has only half the number of chromosomes that the other cells of that organism have (most are diploid). Gametes (like the sperm and the egg) are haploid.

HARDWOOD. Hardwood trees are trees whose wood is dense, strong, and relativel non-absorbent of water. These slow-growing trees include oak and maple. Hardwood trees have broad leaves (and not needles).

HARDY. A hardy plant is one that can withstand the extremes of climate, like frost.

HAUSTORIAL ROOT. An haustorial root is one that absorbs water and nutrients from another plant (and not the soil).

HEATH. A heath is an area of open land that is covered with low-growing shrubs like heather.

HECTARE. A hectare is metric unit of area. A hectare is equal to 10,000 square meters.

HERB. An herb is a seed plant that does not have a woody stem. Every year, herbaceous plants produce a completely new stem. Herbaceous plants are generally short lived and relatively short (compared to woody plants). Some herbaceous plants include the banana, grasses, and forbs.

HERBICIDE. An herbicide is a chemical that kills certain plants. Herbicides are usually intended to kill weeds.

HERB LAYER. The herb layer is another name for the floor of a rainforest.

HERBIVORE. Herbivores (also called primary consumers) are animals that eat plants.

HERMAPHRODITE. A hermaphrodite is a plant that has both female and male reproductive organs.

HETEROTROPH. (pronounced HET-er-oh-TROFE) A heterotroph (also called a consumer) is a living thing that eats other living things to survive. It cannot make its own food (unlike plants, which are autotrophs, making their own food). Animals are heterotrophs.

HETEROPHYLLOUS. A heterophyllous (meaning "other leaf") plant or branch has two or more different leaf shapes on it. Although these leaves have different shapes, they have similar internal anatomical organizations

HILUM. The hilum is the scar on a seed coat at the location where it was attached to the plant's stalk during development.

HOLDFAST. A holdfast is the root-like structure of kelp that holds the base of the plant in place on the sea floor. Unlike real roots, however, the holdfast does not obtain and supply nutrients to the rest of the plant.

HOMOLOGY. Homology is the similarity of characters found in different species that are due to common descent. Examples include the flippers of whales and our arms.

HOMOPLASY. Homoplasy is the similarity of characters found in different species that are NOT from common descent. Examples include the wings of insects and the wings of pterosaurs. These characters derive from convergent evolution, parallel evolution, or character reversal.

HORMONE. A hormone is a chemical in plants (and other organisms) that regulates the plant's growth, reproduction, and other functions.

HORSETAIL.

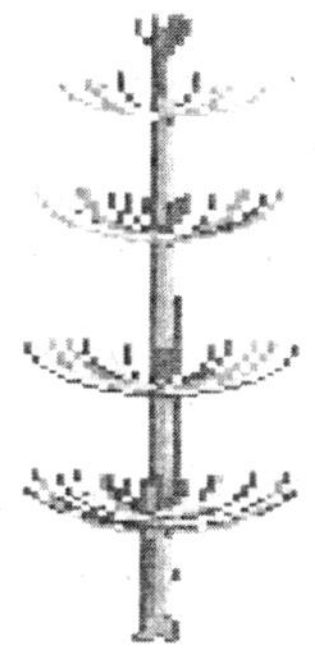

Horsetail is a primitive, spore-bearing plant (a sphenopsid and a pteridophyte) with rhizomes that was common during the Mesozoic Era. Its side branches are arranges in rings along the hollow stem. Horsetails date from the Devonian period 408-360 million years ago, but are still around today and are invasive weeds. Classification: Division Equisetophyta, Class Equisetopsida, Order Equisetales, Family Equisetaceae (Horsetails).

HORTICULTURE. Horticulture is the science of growing fruits, vegetables, flowers, and ornamental plants.

HOST A host is an organism which a parasite uses for food and/or shelter.

HUMUS Humus is the rich, organic portion of the soil. It is composed of decayed plant and animal materials.

HUXLEY, THOMAS H.

Thomas H. Huxley (1825-1895) was a British scientist and friend of Charles Darwin. He was the first scientist to notice the similarity between birds and dinosaurs. He named: Acanthopholis (1865), the family Archaeopterygliдae (1871), Euskelosaurus (1866), Hypsilophodon (1869), and the family Megalosauridae (1869).

HYBRID. A hybrid is the offspring of two organisms that belong to different breeds, varieties, species or genera.

HYDRIC. Hydric environmental conditions are ones that are very wet. Compare with very dry conditions (xeric) and medium moisture conditions (mesic).

HYDROLOGIC CYCLE.

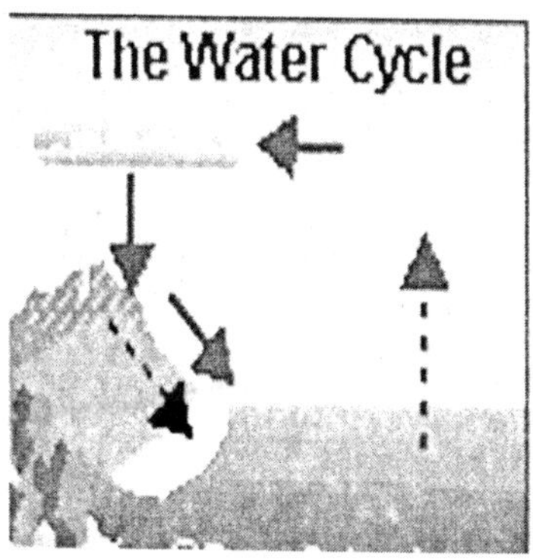

The Hydrologic Cycle (also known as the water cycle) is the journey water takes as it circulates from the Earth to the sky and back again.

HYDROPHYTE. A hydrophyte is a plant that grows in water or in water-logged soil. Hydrophytes have a reduced root system, reduced support and vascular systems, and specialized leaves. Some hydrophytic plants include waterliles and Wolffia (which is the smallest flowering plant). Anchored hydrophytes have a rooting system that is embedded in the soil and they often have floating leaves. (Compare with mesophytes and xerophytes.)

HYDROPONICS. Hydroponics is a method of growing plants in water that contains dissolved nutrients (instead of in soi).

HYOPOCOTYL. The hypocotyl is the part of the stem of a sprouting plant that is above the root and below the stalk of the cotyledons (seed leaves).

I

ICE AGE.

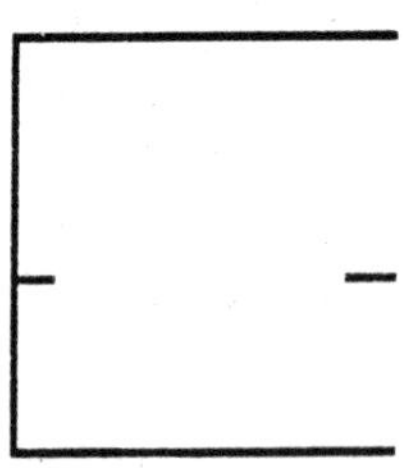

An ice age is a time lasting thousands of years during which the Earth is very cold and largely covered by ice and glaciers.

ICHNITES.

(pronounced IK-nites) Also known as trace fossils or ichnofossils, these are fossilized footprints, nests, dung, gastroliths, burrows, stomach contents, etc., but not actual body parts. Ichnofossils record the movement and behavior of animals. Ichnology is the study of ichnites.

ICHNOFOSSILS. Also known as trace fossils or ichnites, these are fossilized footprints, nests, dung, gastroliths, burrows, stomach contents, etc., but not actual body parts. Ichnofossils record the movement and behavior of animals. Ichnology is the study of ichnofossils.

IGNEOUS ROCK. When molten rock cools, igneous rock is formed.

IMBIBITION. Imbibition is the process in which a seed takes up water from its surrounding and swells. This powerful process precedes germination and can spit the seed coat.

IMPACT CRATER.

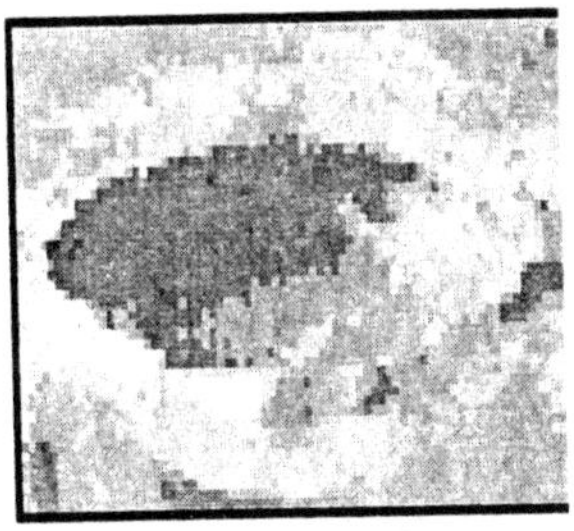

Impact craters are the remains of collisions between an asteroid, comet, or meteorite and the Earth.

IMPERFECT FLOWER.

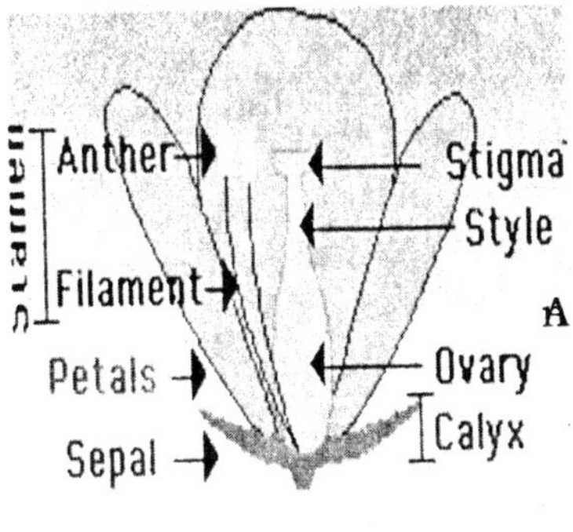

imperfect flower has either male (stamen) and female (ovary) reproductive organs on the same flower, but not both. (Compare with perfect flowers.)

INCOMPLETE FLOWER. An incomplete flower is missing one of the four major parts of the flower, the stamen, pistil, petals, or sepals.

INDEHISCENT. A fruit that remains closed at maturity is indehiscent. Nuts, acorns, sunflowers, grains, and maple pods are examples of indehiscent fruits.

INDEX FOSSILS.

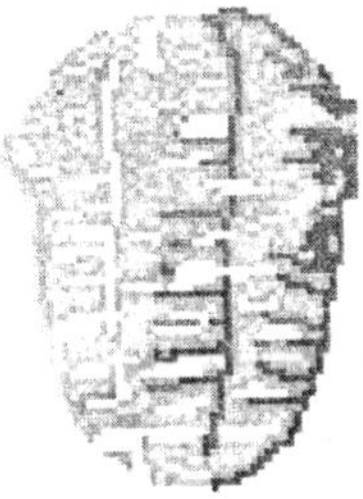

Index fossils are commonly found fossils that are limited in time span. They help in dating other fossils. For example: trilobites were common during the Paleozoic, but not found before the Permian period, and ammonites were common during the Mesozoic Era, but not found after the Cretaceous period.

Another example: the oldest-known ostracods are from the Cambrian period; they became widespread during the Ordovician and remain so.

INDIGENOUS. An indigenous organism is one that lives naturally in a particular region and were not introduced there by man.

INFERIOR OVARY. An inferior ovary (also called an epigynous ovary) is an ovary located below the flower parts (the calyx, corolla, and androecium). The flower parts are attached to the top of the ovary.

INFLORESCENCE.

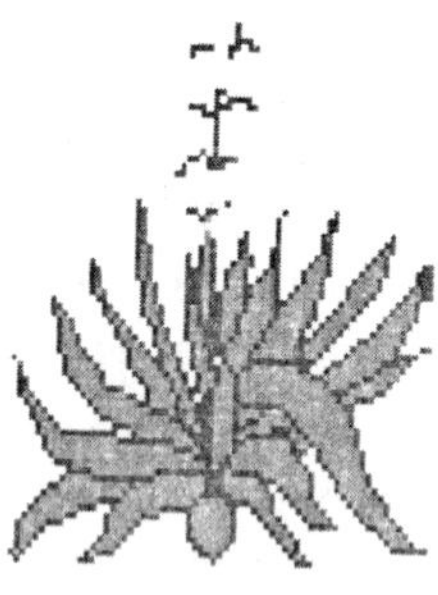

.Inflorescence is the a type of flower in which there is more that one flower in a single structure.

INSECTIVOROUS. Insectivorous organisms eat insects. Insectivorous plants trap and digest insects for nourishment.

INSECTS.

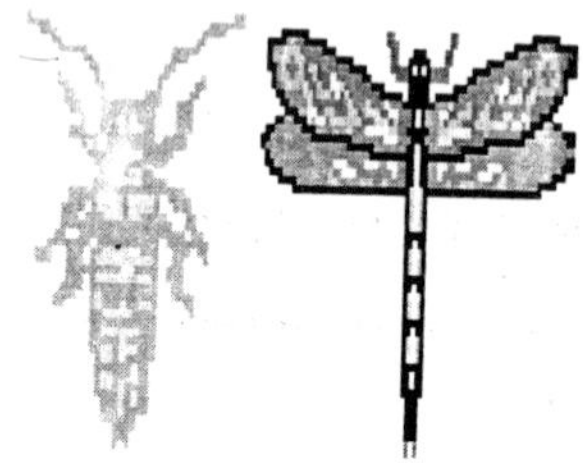

Insects have exoskeletons and six legs. They evolved during the Silurian Period, 438 to 408 million years ago, long before dinosaurs existed.

INTERNODE.

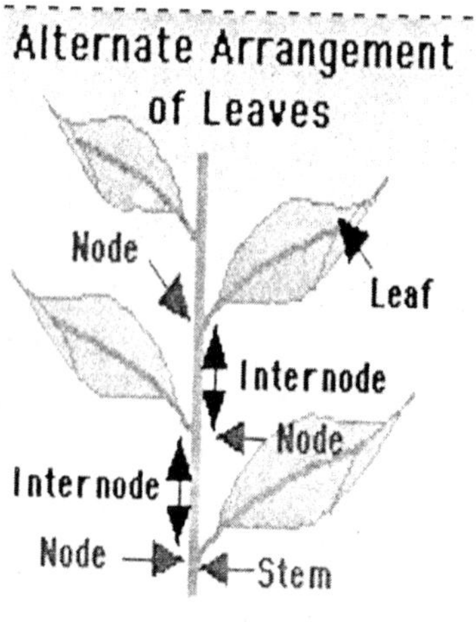

The part of a plant's stem between two nodes is the internode - the distance along the stem between the leaves.

INTERSPECIFIC COMPETITION. Interspecific competition is competition between two different species of organisms (competing for food, water, territory, etc.).

INTRASPECIFIC COMPETITION. Intraspecific competition is competition among members of a species (competing for food, water, territory, etc.).

IRIDIUM.(pronounced irr-ID-ee-um) Iridium is a heavy metal element that is rare on the Earth's surface, but abundant on chondritic meteors and in the Earth's core.

IRIDIUM ANOMALY. The iridium anomaly is a layer of Earth's crust (the K-T layer, which is about 65 million years old) in which there is excess of iridium (a relatively rare element). The presence of this extra Iridium supports the Alvarez asteroid theory, since this iridium may have come from an asteroid.

IRRADIANCE. Irradiance is the amout of light energy from the Sun that hits the Earth in an area.

ISOTOPE. An isotope of an element is another form of the same element that has a different number of neutrons in the nucleus (giving it a different atomic weight).

J

JUNGLE. A jungle is a rainforest or a forest whose canopy trees have been logged (cut), causing lush growth on the forest floor (also known as a secondary forest).

JUNIPER. Junier is a conifer and a gymnosperm with blue, woody fruits and overlapping, scale-like leaves. Classification: division Coniferophyta, family Cupressaceae.

JURASSIC PERIOD.

Birds evolved and many dinosaurs flourished during the Jurassic period, 208-146 million years ago. Flowering plants appeared about 140

million years ago, during the late Jurassic period.

JUVENILE .A juvenile plant is is in an early phase of plant growth in which it increases in size but has not yet flowered.

K

KAPOK TREE.

The kapok tree, *Ceiba pentandra* or *Eriodendron anfractuosum*, is a large, deciduous, tropical tree that is native to tropical America, Africa, and the East Indies. This fast-growing tree is generally from 45 to over 100 feet (14-30 m) tall; the kapok is the tallest tree in Africa. It has pink, white, or yellow flowers in clusters. The lightweight silky down from the seed pods (sometimes called Java cotton) is used as pillow stuffing, sleeping bag stuffing, life jacket stuffing, furniture upholstery, insulation, and for other uses. The green leaves are lanceolate (lance shaped) and palmately compound (with 5 to 9 leaflets). The yellow-green oil from the seeds is used in foods and to manufacture soap. Young leaves are also cooked and eaten; the wood is also used. The night-blooming flowers are pollinated and the seeds are spread by fruit bats.

KELP. Kelp is a cold-water seaweed that belongs to the family Laminariaceae. It is a large, floating plant that grows in large masses (called kelp forests) and is held to the sea floor by a holdfast. Kelp is usually brown. This seaweed is used by people as food, a food additive, fertilizer (when reduced to ash), and as an ingredient in many other products. Kelp evolved during in the Miocene epoch.

Kelp is included in either kingdom Protista or Chromista.

KELP FOREST. A kelp forest is a marine ecosystem (a biome) that is dominated by large kelp plants. These oceanic forests grow in cold and temperate waters, especially along the western coasts of the continents.

KINGDOM. In classification, a kingdom is the highest grouping of similar organisms. A kingdom contains one or more phyla (plural of phylum). Life on Earth is divided into five kingdoms: Animalia (animals), Plantae (plants), Fungi, Protista (protozoans and eucaryotic algae), and Monera (blue-green algae).

K-T BOUNDARY. The K-T Boundary was the boundary between the Cretaceous and Tertiary periods, occurring 65 million years ago. This was a time of the huge K-T mass extinction.

K-T EXTINCTION. The K-T extinction was the mass extinction that occurred 65 million years ago, at the boundary of the Cretaceous and Tertiary periods.

KUDZU. Kudzu is a type of legume, a flowering plant that bears its protein-rich seeds in pods and can fix nitrogen from the soil (due to the symbiotic root bacteria rhizobia). Its tubers are edible and the fiber is useful. The plant has a woody stem, wide leaves, and purple flowers. Kudzu is native to Japan and was introduced to the United States around 1876; it soon became a nuisance weed in the southern US. Classification: Kingdom Plantae, class Magnoliopsida, order Rosales, family Leguminosae (pulses), genus Pueraria, specie *P. hunbergiana.*

L

LAGERSTATTEN. Lagerstätten (meaning "fossil deposit places" in German) are geological deposits that are rich with varied, well-preserved fossils, representing a wide variety of life from a particular era. These spectacular fossil deposits represent an amazing "snapshot" in time. Some Lagerstätten include the La Brea Tar Pits (California, USA), Ediacara Hills (South Australia), Burgess Shale (B.C., Canada), Solnhofen (Germany), and Mazon Creek (Illinois, USA)

LAMINA.

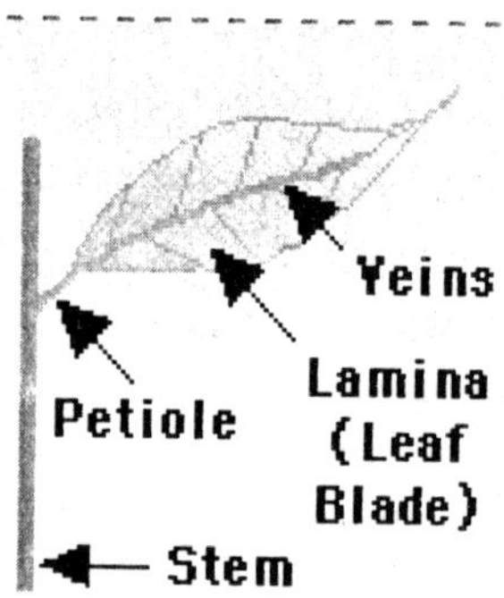

The lamina is the blade of a leaf.

LANCEOLATE.

Lanceolate leaves are shaped like a lance; they have a broad base and taper to a point.

LATERAL. Lateral means of, near, or from the side of an organism.

LATEX. Latex is a milky substance that is tapped from rubber trees.

LAURASIA.

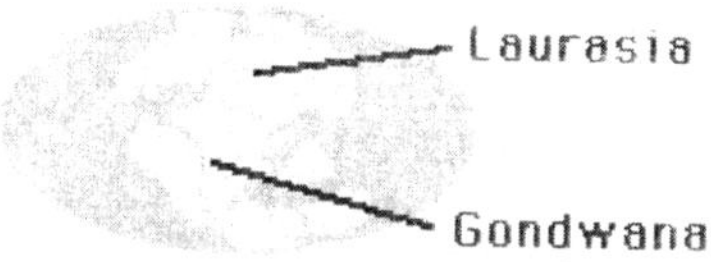

Laurasia was the northern supercontinent formed after Pangaea broke up during the Jurassic period. Laurasia included what are now North America, Europe, Asia, Greenland, and Iceland.

LAVA.

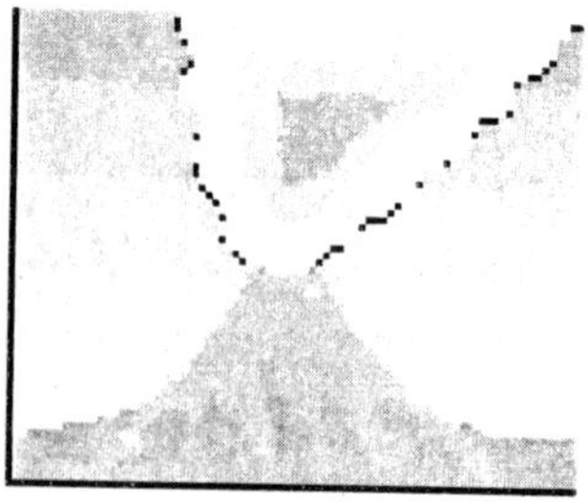

Lava is molten rock. It usually comes out of erupting volcanoes.

LEAF. A leaf is an outgrowth of a plant that grows from a node in the stem. Most leaves are flat and contain chloroplasts; their main function is to make food energy through photosynthesis. The first leaf to grow from a seed is called the cotyledon.

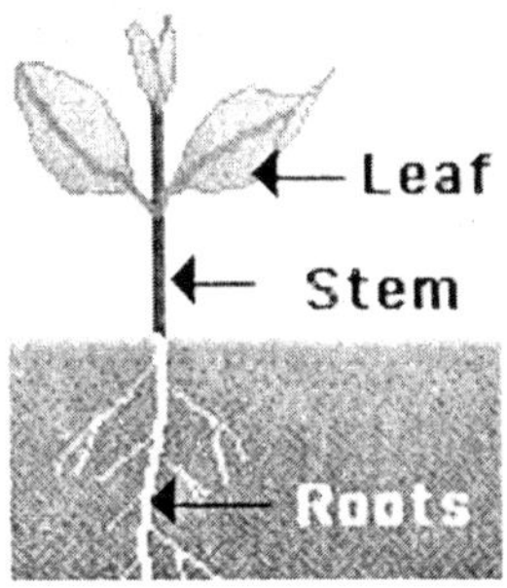

LEAF ABSCISSION. Leaf abscission is the normal separation of a leaf from a plant. Abscisic acid is the plant hormone involved in leaf abscission.

LEAF AXIL.

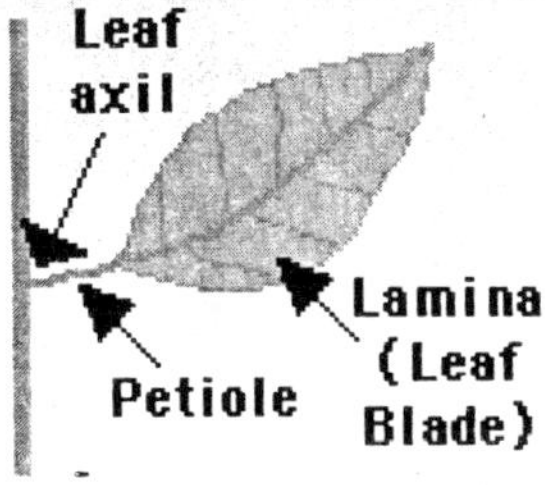

The leaf axil is where the petiole of the leaf attaches to the stem.

LEAF DIMORPHISM. Having two types of leaves on the same plant.

LEAF GAP. A leaf gap is a break or gap in the plant's stele (cylinder of vascular tissue) at the base of a node (where a leaf is). In the absence of a leaf, there would be no gap, i.e., an unbroken cylinder of vascular tissue

LEAFLET.

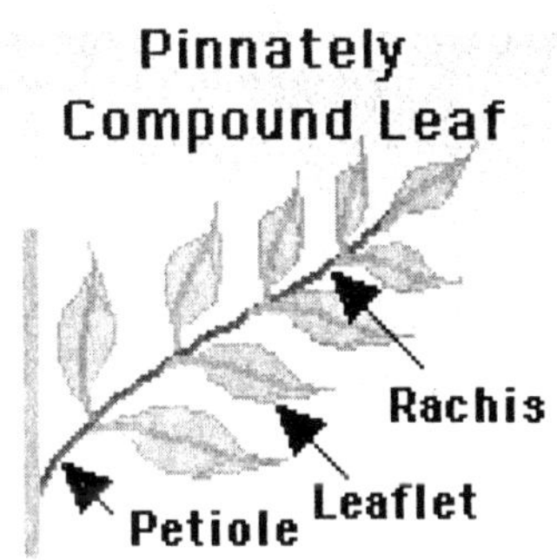

A leaflet is part of a compound leaf. Each leaflet looks like a small leaf.

LEAF SCAR. A leaf scar is mark on a plant's stem that is left where a leaf was once attached.

LEGUME. A legume is a flowering plant that bears its protein-rich seeds in pods and can fix nitrogen from the soil (due to the symbiotic root bacteria, rhizobia). Some legumes include lentils, beans, clover, alfalfa, lespedezas, vetches, kudzu, and peas. Classification: Kingdom Plantae, class Magnoliopsida, order Rosales, family Leguminosae.

LEMMA. A lemma is a bract (a reduced, leaf-like structure associated with a flower) in a grass spikelet that is located below the stamens and pistil of the flower.

LENTICEL. A lenticel is a small pore in the bark of a tree through which gas exchange occurs.

LEPIDODENDRON

Lepidodendron (also known as the "scale tree") was a giant club moss (a tree-like plant) whose long trunk had bark with a diamond-shaped pattern (the scars of old, dead branches that fell off). This ancient lycopod was over 130 ft (40 m) tall; the trunk was over 6 ft (1.8 m) in diameter. Lepidodendron lived in swampy areas during the Carboniferous Period (about 360 to 286 million years ago). It has spirally-arranged leaves that ended in cones. By the time the dinosaurs lived, the giant club mosses had died out and were replaced by smaller club mosses.

LEPTOCYCAS.

Leptocycas gracilis was a cycad (a primitive seed plant) that lived during the late Triassic period. It was a palm-like tree with a long, woody trunk that lived in warm climates. This tree was about 4.8 ft (1.5 m) tall. Separate male and female plants exist (they are dioecious). This gymnosperm had long, divided leaves and produced large seed cones.

LIANA. Liana is a woody, climbing vine that grows on tree trunks in order to reach sunlight in the rainforest. Rattan, used for making wicker furniture, is made from liana vines.

LIAS EPOCH. The Lias epoch was the early part of the Jurassic period, about 206 to 180 million years ago.

LICHEN. Lichens are organisms that are a symbiosis between algae and fungus. The photosynthetic algae produces food and the fungus provides protection for the algae. Lichens can survive in a variety of conditions and are found worldwide.

LIGNOTUBER. A lignotuber is a root or woody stem base that stores water.

LIGNULE.

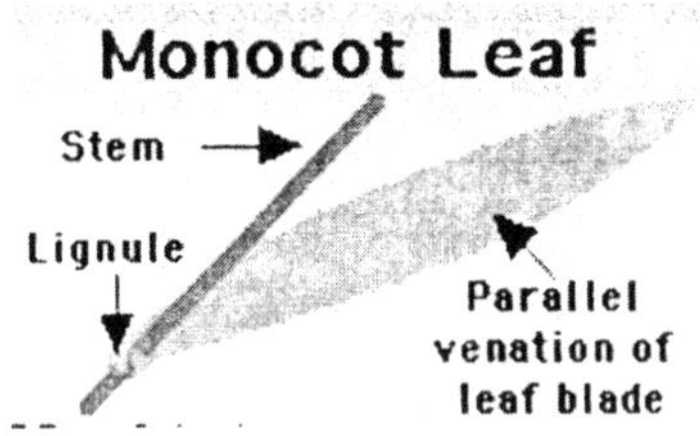

Lignules (which means tongue) are the two small flaps at the base of a monocot's leaf that wrap around the stem.

LIPID. A lipid is a fat. Membranes in cells are composed of two lipid layers (with proteins dissolved within them).

LINNEAN SYSTEM

The Linnean System is a method of classifying organisms based on a simple hierarchical structure. Organisms are divided into groups using the following system: Kingdom, Phylum, Class, Order, Family, Genus and Species.

LIVERWORT. Liverwort is a green, moss-like plant belonging to the family Hepaticae. This small, simple plant lives in moist, shady areas and has wide, flat leaves that lie close to the ground. Liverworts reproduce with spores.

LIVING FOSSILS .Living fossils are organisms that lived during ancient times and still live today, like the Coelacanth, the horseshoe crab, the gingko tree, cycads, horsetails, club mosses, and many, many other well-adapted organisms.

LOAM. Loam is a type of rich, crumbly soil that contains an almost equal amount of sand and silt, plus a smaller amount of clay (it contains from 28 to 50% silt, less than 52% sand, and 7 to 27% clay).

LOBED. A lobed leaf is one in which the margin is divided into rounded or pointed sections and the incisions (cuts) go less than halfway to the midrib.

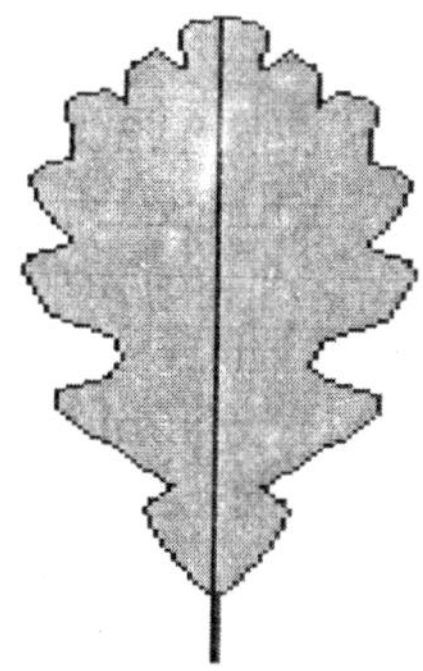

lobed

LOESS. Loess is the term for soil particles that have been moved to another location by the wind.

LONGITUDINAL STRIATE VENATION. Longitudinal Striate venation is a vein pattern found in monocots leaves. A leaf with longitudinal striate venation has its veins arranged almost parallel to one another, running the length of the leaf (also called parallel venation).

LOW SEASONALITY. Low seasonality is when there is a small difference in temperature between the seasons (with mild winters and summers). Compare to stong seasonality, in which the difference in temperatures between the seasons is big (for example, with a hot summer and a cold winter).

LYCOPSIDS. Lycopsids (club mosses) are primitive, vascular plants (pteridophytes) that evolved over 375 million years ago (during the Devonian). Huge club mosses went extinct during the Permian mass extinction; smaller ones lived during the time of the dinosaurs. These plants live near moisture (in order for their spores to germinate). These fast-growing, resilient plants propagate with rhizomes (underground stems).

LYSOSOME. Lysosomes are organelles within plant cells (and other eukaryotic cells) that contain digestive enzymes and is involved with the digestion of food. A lysosome fuses with a vacular membrane that contains food, and the lysosome's enzymes digest the food, breaking the food down into its components.

M

MAASTRICHTIAN AGE. The Maastrichtian age was the last part of the Cretaceous period, about 71 to 65 million years ago, at the very end of the Mesozoic Era.

MAGNETIC FIELD.

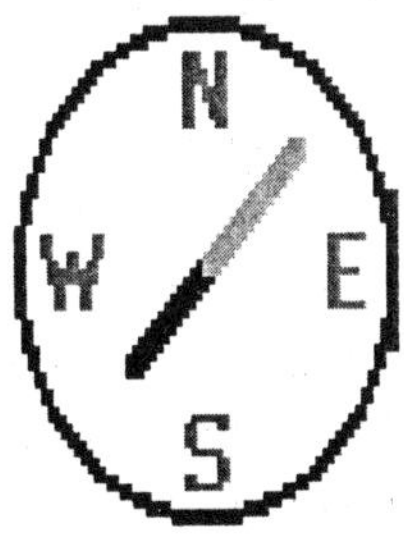

The Earth's magnetic field is aligned with the north and south poles, and has reversed many times during geologic history. A fossil's magnetic orientation can give clues to its date.

MALM EPOCH. The Malm epoch was the late (or upper) part of the Jurassic period, about 159 to 144 million years ago.

MANGROVE. Mangroves are tropical evergreen trees and shrubs. They live near the shore in tropical marshes and tidal shores with their adventitious roots in the salt water. Classification: genus Rhizophora. Some mangroves include the red mangrove (*Rizophora mangle*), the black mangrove (*Avicennia germinans*), and the white mangrove (*Laguncularia racemosa*).

MANGROVE FOREST. A mangrove forest (also called a mangrove swamp) is an expanse of mangrove trees. These trees live near the shore in tropical marshes and tidal shores with their adventitious roots in the salt water.

MARGINS. The margins of a leaf are its edges, which very from plant to plant. The margins can be smooth, serrated, or toothed; they can also be lobed or entire.

MASSARANDUBA TREE. The Massaranduba tree (*Manilkara bidentata)*), also called the cow tree, macaranduba, bulletwood, and balata, is a large rainforest tree whose sap is milky white and is edible. Its fruit is also edible. This tree grows to about 100 to 150 feet tall and has a diameter of 2 to 6 feet. The tree's unusual bark is deeply-scored, reddish and ragged. The bark is used to produce a red dye. Many Massaranduba trees are in the rainforest are being harvested for its durable lumber.

MASS EXTINCTION. Mass extinction is the process in which huge numbers of species die out suddenly. The dinosaurs (and many other species) went extinct during the K-T extinction, probably because of an asteroid that hit the Earth.

MEDULLARY RAYS. Medullary rays (also called wood rays) are bundles of cells that radiate from the center of a tree like the spokes of a wheel (they are most easily seen in the cross section of a tree trunk). They store food and transport it horizontally within the tree.

MEIOSIS. Meiosis is a process in which an eukaryotic cell's (diploid) nucleus divides twice and produces four daughter cells (called gametes or reproductive cells), each of which have half the genetic complement of the parent cell (they are haploid). This process was named by Farmer and Moore in 1905.

MERISTEM. The meristem is a group of plant cells that can divide indefinitely. The meristem provides new cells for the plant.

MESIC. Mesic environmental conditions are ones with medium moisture (water). Compare with very dry conditions (xeric) and very wet conditions (hydric).

MESIC SOIL. Mesic soil is a medium type of soil (hence the name, which means middle) that drains well yet retains some water. Mesic soil is exellent for farming.

MESOPHYLL. Mesophyll is the internal tissue of a photosynthetic leaf.

MESOPHYTE. A mesophyte is a plant that has moderate water requirements. (Compare with hydrophytes and xerophytes.)

MESOZOIC ERA.

The Mesozoic Era ("The Age of Reptiles"), occurred from 248-65 million years ago. It is divided into the Triassic, Jurassic, and Cretaceous periods. Dinosaurs, mammals, and flowering plants evolved during the Mesozoic, and Pangaea broke up. The era ended with the K-T mass extinction.

METAMORPHIC ROCK

Metamorphic rocks are compacted by pressure and heat from deep inside the earth.

METEOR.A meteor is a meteoroid that has entered the Earth's atmosphere, usually making a fiery trail as it falls. It is sometimes called a shooting star. Most burn up before hitting the Earth.

METEORITE. A meteorite is a meteor that has fallen to Earth. Meteorites are either stone, iron, or stony-iron.

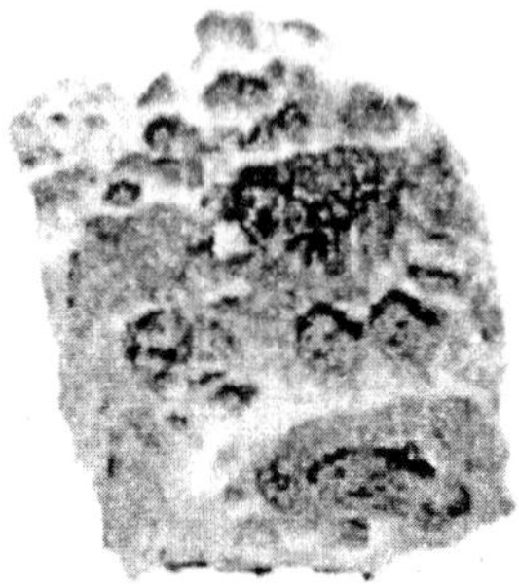

METEOROID. Meteoroids are tiny stones or pieces of metal that travel through space.

MICROPYLE. The micropyle is the small pore in a seed that that allows water absorption.

MIDRIB. The midrib (rachis) is the central rib of a leaf. It is usually continuous with the petiole and is often raised above the lamina (the leaf blade). On a compound leaf, the midrib extends from the first set of leaflets (where the petiole ends) to the end of the leaf.

MILANKOVITCH THEORY. The Milankovitch Theory attempts to explain major temperatures changes on Earth, especially ice ages, by slight variations in the amount of sunlight reaching the Earth caused by the eccentricity of the Earth's orbit around the Sun. The Earth's orbital eccentricity changes the Earth's average distance from the sun and therefore slightly changes the amount of sunlight reaching the Earth. Milankovitch looked over the past 600,000 years and correlated summer temperature mimnima with four major ice ages in this time. Eccentricity cycles last over 100,000 years. This theory was proposed by Milutin Milankovitch in 1938. Recently, scientists (Richard A. Muller and Gordon J. MacDonald, July 11, 1997, *Science*) have found that changes in the axal tilt of the Earth's orbit more closely match glacial cycles for the past million years.

MILDEW. Mildew is a parasitic, filamentous fungus that grows on a host plant.

MILKWEED. Milkweed (genus Asclepius) is a common plant that contains toxins (poisons). There are more than 100 species of this perennial herb, containing varying concentrations of toxic chemicals (glycosides). Monarch butterflies (and other milkweed butterflies) lay their eggs on milkweed leaves; the caterpillars eat milkweed leaves to incorporate the milkweed toxins into their bodies in order to poison their predators.

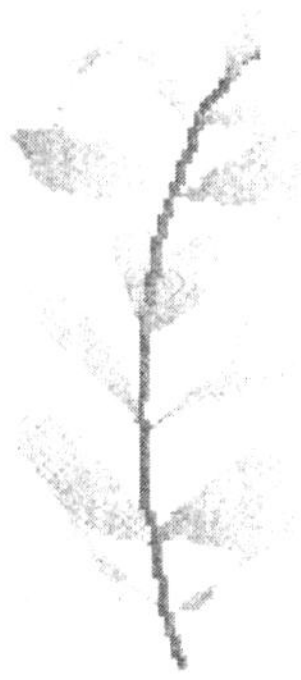

MILKWEED BUTTERFLIES.

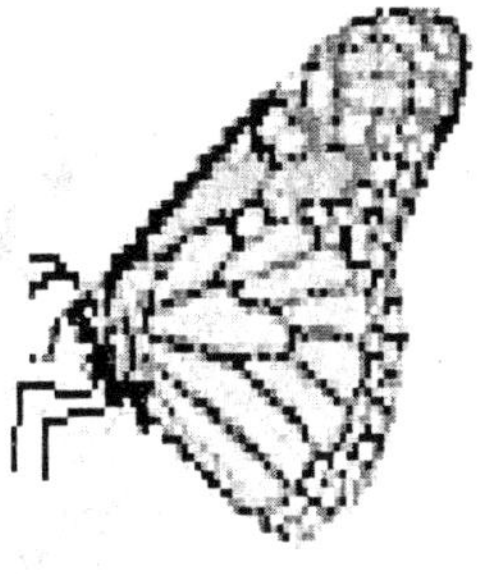

Milkweed butterflies are members of the family Danaidae. As larva, they eat the milkweed plant. Milkweed butterflies include the Monarch (*Danaus plexippus*), the Queen (*Danaus gilippus*), and others.

MILLION. A million is a thousand thousand. The dinosaurs lived millions of years ago.

MINERAL. A mineral is a naturally-occurring solid of definite chemical composition whose atoms usually form a regular pattern.

MITOCHONDRION. Mitochondria (the plural of mitochondrion) are organelles within most eukaryotic cells - they power cells by generating energy (in the form of ATP, adenosine triphosphate) by breaking down glucose (a type of sugar). Mitochondria are self-replicating and bound by membranes.

MITOSIS.

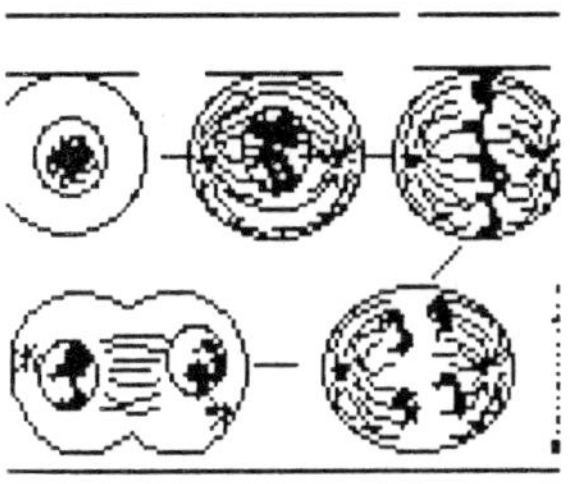

Mitosis is a process in which an eukaryotic cell's nucleus

divides after having duplicated its chromosomes - it produces two daughter cells that are genetically identical to the parent cell and to each other. Walter Flemming (1843-1905), discovered and named this process in 1870.

MIXED WOODLAND. A mixed woodland is a woodland that has both conifers and broadleaf trees.

MOLD. Mold is a type of fungus (and not a plant). Like other fungi, molds do not contain any chlorophyll (and cannot make their own food); molds live off the food produced by plants or animals, or decaying matter. Molds are often parasites on plants, animals, or even other fungi. Molds reproduce with spores. Some molds spoil our food, but other foods are produced by the action of mold (for example, blue, Roquefort, and Camembert chesses have mold growing in them, giving them their flavor). The antibacterial drug penicillin is made from the Penicillium mold (Alexander Fleming discovered penicillin in 1928). Classification: Kingdom Fungi, Division Eumycota (septate fungi), Classes Hyphomycetes, Oomycetes, and Zygomycetes.

MONERA. Monera is a prokaryotic kingdom (separate from the plant kingdom) that includes the earliest forms of life on Earth, like archaebacteria (the oldest types of bacteria), eubacteria (like *E. coli*), and cyano-bacteria (blue-green bacteria).

MONOCOT. A monocot is a type of flowering plant (an angiosperm) whose seed has one embryonic leaf (cotyledon). The leaves of monocots generally have parallel venation (the veins are parallel to one another). The roots of monocots are usually fibrous and the flower parts are often in multiples of three.

MONOCULTURE. Monoculture is a system of agriculture in which a single type of crop is grown in an area.

MONOECIOUS. Monoecious plants have the male and female reproductive organs on the same plant. (Compare with dioecious.)

MONOPHYLETIC. A monophyletic group consists of all organisms that share a particular common ancestor (and therefore have similar features). The members of a monophyletic group are more closely related to one other than they are to any organism outside the group. A monophyletic group is also called a clade. An example of a monophyletic group is mammals.

MONOPODIAL. Monopodial (meaning "one foot") is a type of plant growth in which the plant has a single main stem. Compare with sympodial

MORPHOLOGY. Morphology is the study of the external structure of organisms (for example, the arrangement of leaves on a plant).

MORRISON FORMATION. The Morrison Formation is a rock outcropping located in Utah, USA. This exposed sedimentary rock dates from the late Jurassic period, when this area was similar to a savanna (without the grass, since flowering plants hadn't evolved yet). Dinosaurs like Allosaurus, Seismosaurus, Ceratosaurus, Ornitholestes, and Diplodocus have been found at the Morrison foundation.

MOSS. Moss is a small, low-growing, green plant that has a stem, leaves, and rhizomes (but no vascular system). Mosses reproduce with spores or by forming gametes. Classification: Kingdom Plantae (plants), Plylum Bryophyta (mosses), Four Classes: Class Sphagnopsida (Peat Mosses), Class Takakiopsida ("Takakiophytes"), Class Andreaeopsida (Granite Mosses), Class: Bryopsida or Musci ("True" Mosses). The true mosses (Class Bryopsida/Musci) are divided into three subclasses: Subclass Polytrichidae, Subclass Tetraphidae, Subclass Bryidae.

MUCIGEL. Mucigel is a slimy, protective substance that is secreted by plant roots.

MULTICELLULAR. Multicellular organisms consist of many cells. (Compare with unicellular.)

MUSHROOM.

Mushrooms are fast-growing fungi (they are not plants). They grow in dark, damp places and reproduce via spores.

MUTUALISM. Mutualism is an association between two different species of organisms in which both benefit from the association. An example of mutualism is the relationship between bees and the flowers they sip nectar from-the bees get nectar from the flower and the flower gets pollinated by the bees. Mutualism used to be called symbiosis (which is now a more general term).

MYA. "mya" stands for millions of years ago.

MYCORRHIZAE. Mycorrhizae is a fungus that grows in a symbiotic relationship with the roots (or rhizoids) of a plant.

N

NANOFOSSIL. Nanofossils are microscopic fossils that are very abundant, widely distributed, and time-specific (because of their high evolutionary rates). They are very useful index fossils.

NATURAL SELECTION. Natural selection is the process in which some organisms live and reproduce and others die before reproducing. Some life forms survive and reproduce because they are better suited to environmental pressures, ensuring that their genes are perpetuated in the gene pool.

NATURALIST. A naturalist is a person who studies plants and animals.

NECTAR. Nectar is the sweet liquid produced by many flowers. Nectar attracts many insects (like butterflies and bees) who go from flower to flower sipping nectar, causing the pollination of the flowers.

NEEDLE. A needle is a long, thin, pointed, needle-shaped leaf, like that of the pine. The word acicular is sometimes used to describe needles.

NEOCOMIAN EPOCH. The Neocomian epoch was the early (lower) part of the Cretaceous period, about 144 to 127 million years ago.

NEOGENE

The Neogene (24 million to 1.8 million years ago) was the later part of the Tertiary Period. It is divided into the Miocene Epoch (24 million to 5 million years ago, when many mammals appeared, including the horses, dogs, bears, South American monkeys, apes in southern Europe, and Ramapithecus; also, modern birds appear) and the Pliocene Epoch (5 million to 1.8 million years ago, when the first hominids (australopithecines) developed, modern forms of whales appeared, and Megalodon swam the seas).

NEOTROPICAL. The neotropical region extends from the southern tip of South America up to Mexico. This vast region includes many different habitats, including tropical rainforests.

NICHE. A niche is a place or position that an organism occupies in an environment.

NITROGEN FIXING. Some bacteria (rhizobia) are nitrogen-fixing; they transform nitrogen gas in the atmopshere into a form that can be used by plants.

NODE.

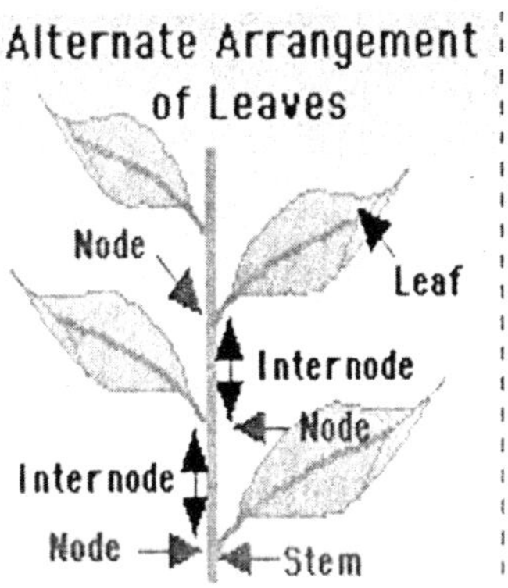

A node is a part of the stem of a plant from which a leaf, branch, or aerial root grows.

NODULE. A nodule is a small, rounded knot on some leguminous plant roots which contain nitrogen-fixing bacteria (rhizobia). For example, peanut plant roots have nodules.

NOXIOUS WEED. A noxious weed is a plant that is considered by local authorities to be a problem, growing where it is not wanted.

NUCLEAR ENVELOPE. (also called the nuclear membrane) The nuclear envelope is double membrane that surrounds the nucleus.

NUCLEUS. The nucleus is a membrane-bound organelle in each plant cell which contains the genetic material of the cell (DNA in chromosomes). It is where DNA (deoxyribonucleic acid) replicates itself, and where RNA (ribonucleic acid) is made. The plural of nucleus is nuclei.

NUCLEOLUS. The nucleolus is a spherical or ovoid organelle in the nucleus of a eukaryotic cell. It consists of granules of DNA and RNA, and synthesizes ribosomal RNA. The plural of nucleolus is nucleoli.

NURSE LOG.

A nurse log is a fallen tree trunk that nourishes new life on the forest floor as it decomposes. It provides a rich, moist habitat for ferns, lichens, mosses, and tree seedlings.

NUTRIENT. A nutrient is a chemical that an organism need to ingest in order to survive (like fats, carbohydrates, vitamins, minerals, etc.).

NUCLEAR MEMBRANE. (also called the nuclear envelope) The nuclear membrane is double membrane that surrounds the nucleus of the cell.

O

OMNIVORE.

Omnivores are animals that eat both animals and plants.

ORDER. In Linnean classification, an order is a group of related or similar organisms. An order contains one or more families. A group of similar orders forms a class.

ORDOVICIAN PERIOD. (pronounced or-da-VISH-ee-en) The Ordovician Period lasted from 505 to 438 million years ago. Primitive plants appear on land, as do the first corals, primitive fishes, seaweed and fungi. Graptolites, bryozoans, gastropods, bivalves, and echinoids also appear. There were high sea levels at first, then global cooling and glaciation, and extensive volcanism. North America was under shallow seas. The Ordovician ended in huge extinction, probably due to glaciation. The Ordovician was named for an ancient tribe (the Ordovices) that lived in northern Wales.

ORGAN. An organ is a specialized tissue (a group of cells) in multicellular organisms that performs a particular function. For example, the ovary is an organ that produces ovules.

ORGANELLE. An organelle is membrane-bound structure within a plant's cell (and all eukaryotic cells) where specialized metabolic tasks occur. Some organelles include the nucleus, nucleolus, mitochondria, the ER (endoplamic reticulum), and lysosomes.

OSMOSIS. Osmosis is a process in which water moves across a semi-permeable membrane in order to equalize the concentration of the solution on both sides. For example, water will move across a cell membrane from areas of low salinity (a hypotonic solution) to areas of high salinity (a hypertonic solution) to equalize the concentrations (and create two isotonic solutions).

OSMOTIC PRESSURE. Osmotic pressure is a phenomenon in which water exerts pressure against a membrane (when it is bordered by solutions of differing concentrations).

OSMOTIC POTENTIAL. Osmotic potential is the potential of water to move into a region by the process of osmosis, the potential of the water to travel from a hypotonic (low concentration) solution to a hypertonic (high concentration) solution.

OUTCROPPING. An outcropping is a place where the bedrock (the underlying rock) is exposed on the Earth's surface.

OVARY,

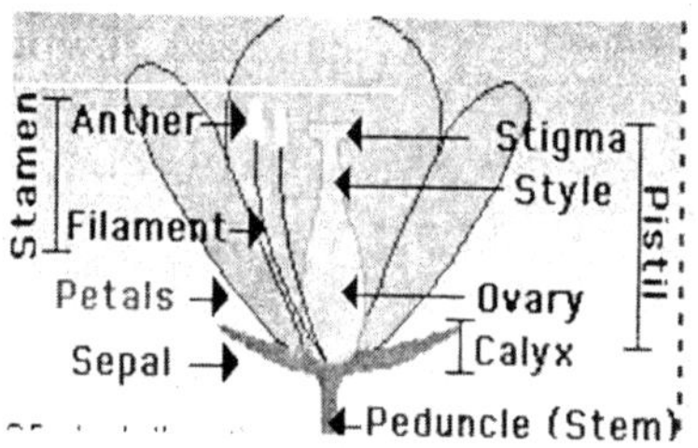

An ovary is a female reproductive organ in plants that produces ovules.

OVERBURDEN. The overburden is the rock that lies on top of a fossil.

OVULE. An ovule is the female reproductive cell of flowering plants and cone-bearing plants. After the ovule is fertilized by the male pollen, the ovule becomes a seed.

P

PALEOBOTANY. Paleobotany is the branch of botany that studies the plants that existed in former geologic periods, chiefly by studying fossils.

PALEONTOLOGY. Paleontology is the branch of biology that studies the forms of life that existed in former geologic periods, chiefly by studying fossils. "Paleo" means old or ancient. "Ontology" is the study of existence ("onto-" means existence, "-logy" is the study of something).

PALEONTOLOGIST. A paleontologist is a scientist who studies paleontology, the forms of life that existed in former geologic periods, chiefly by studying fossils. "Paleo" means old or ancient. "Ontology" is the study of existence ("onto-" means existence, "-logy" is the study of something). This "-ist" at the end means a person who is involved in the field.

PALEOZOIC ERA. The Paleozoic Era (540 to 248 million years ago) saw an explosion of new life forms. The Paleozoic (meaning "ancient life") ended with the largest mass extinction in geologic history and was followed by the Mesozoic Era. It is divided into the Cambrian, Ordovician, Silurian, Devonian, Carboniferous, and Permian periods. It was followed by the Mesozoic Era, the time of the dinosaurs.

PALISADE PARENCHYMA. The palisade parenchyma (also called the palisade layer) is a layer of tightly-packed, tube-shaped, chlorophyll-containing cells (chlorenchyma) that is the major photosynthetic layer of a leaf. The palisade layer is located in the upper mesophyll of a leaf (just beneath the upper dermal layer of the leaf).

PALM. A palm is an evergreen tree and a monocot. Classification: Division Magnoliophyta (Angioperms), Class Liliopsida (Monocots) , Subclass Arecidae, Order Arecales, Family Arecaceae (or Palmae) (Palms).

PALMATE.

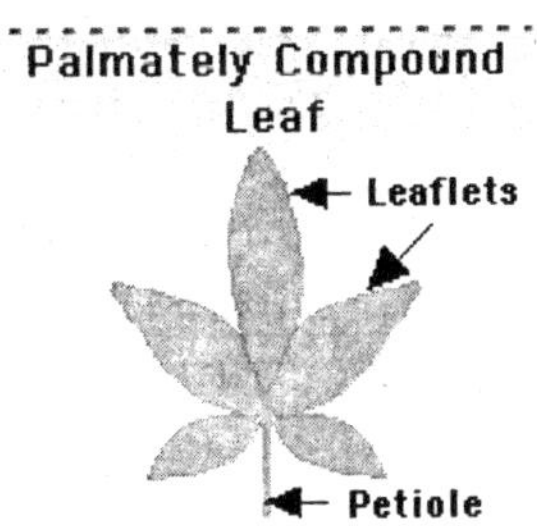

A palmate leaf has a hand-like structure. It has more than three lobes that branch from a single point at the base of the leaf.

PALMATE VENATION

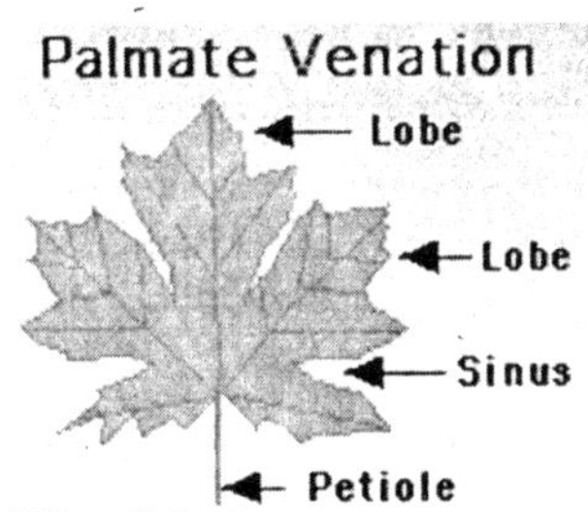

A leaf with palmate venation has the main veins arising from a point at the base of the leaf.

PANGAEA. (pronounced pan-GEE-ah) Pangaea was a supercontinent consisting of all of Earth's land masses. It existed during the Permian through the Jurassic period. It began breaking up during the Jurassic, forming the continents Gondwanaland and Laurasia.

PANTHALASSA. (pronounced pan-tha-LASS-ah) Panthalassa (meaning "All seas") was the super-ocean that existed on Earth during the time of the super-continent Pangaea. Panthalassa existed during the Permian through the Jurassic period, when Pangaea began to break up; the Tethys sea formed between the northern and southern parts of pangaea as they drifted apart

PARAPHYLETIC. A paraphyletic group (also called a grade) consists of a common ancestor and some, but not all, of its descendants. These are incomplete groups based primarily on physical characteristics rather than directly on evolutionary relationships. An example of a paraphyletic group is the dinosaurs (without including the birds)

PARALLEL VENATION.

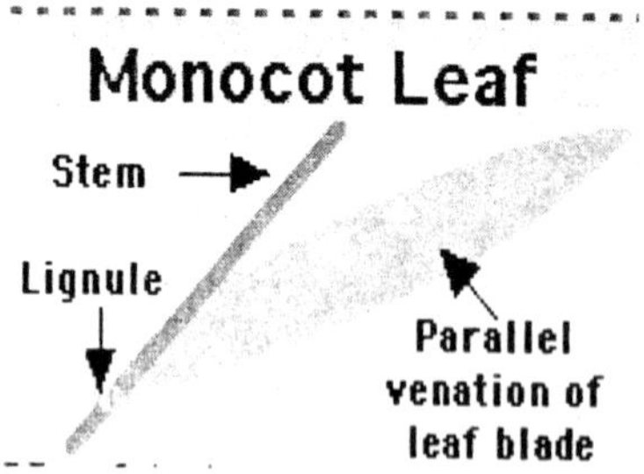

A leaf with parallel venation (also called longitudinal-striate venation) has its veins arranged almost parallel to one another.

PARASITE. A parasite is an organism which uses its host for food and/or shelter. A parasite gives its host nothing in return, and often makes it sick or even kills the host. Termites are a parasite of many trees.

PARASITISM. Parasitism is arelationship between two organisms in which one organism benefits at the other organism's expense. Lice are an example of a parasite that affects many animals; termites are a parasite that are destructive to many trees. Parasitism is a type of symbiosis.

PARENCHYMA. Parenchyma are generalized (undifferentiated) cell or tissue in a plant. Parenchyma cells make or store food; they can often divide or differentiate into different types of cells and have thin cell walls. Parenchyma is the most common type of plant cell. The pith is parenchyma cells at the center of the primary stem of a dicot.

PARSIMONY. Parsimony is the scientific idea that the

simplest explanation of a phenomenon is the best one.

PARTED.

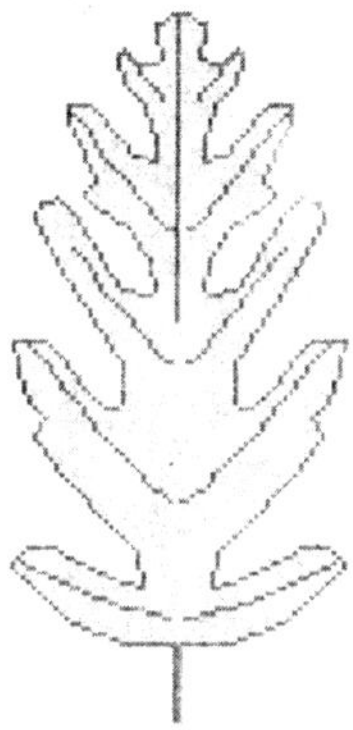

parted

A parted (also called cleft) leaf is one in which the margins between the irregular teeth go more than halfway to the midrib.

PEANUT PLANT

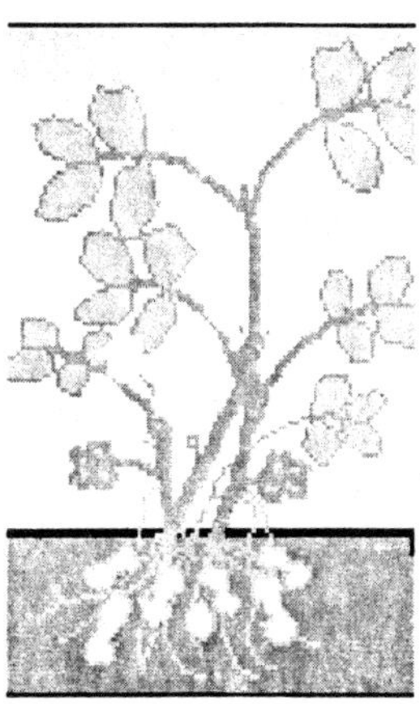

The peanut plant (*Arachis hypogea*) is a legume that is native to South America (it now grows around the world). The peanut plant grow to about 60 cm tall and has yellow flowers (1-2 cm long). The flower blooms for about half a day. About 4 days later, a stem (also called a peg) will grow from the flower and head into the soil. At the end of each stem, the seed pods (peanuts in the shell) will grow. The peanut is an annual plant (it completes its life cycle in one year).

PEAT. Peat is a type of soil that is composed of incompletely decomposed plant material that water-logged and low in oxygen.

PEDICEL. A pedicel is plant stalk that attaches a single flower or fruit to the main branch of the inflorescence.

PEDOGENESIS. Pedogenesis is the natural process in which soil forms.

PEDUNCLE. A peduncle is plant stalk that bears an inflorescence or single flower.

PENNSYLVANIAN PERIOD. The Pennsylvanian Period lasted from 325 to 280 million years ago. During this time, the first reptiles (like Hylonomus) appeared and ferns dominated the warm, swampy landscape.

PERENNATE. A plant that perennates lives from year to year (it is a perennial).

PERENNIAL. A perennial plant has a life cycle that lives for more than two years. Perennials usually flower each year.

PERFECT FLOWER .A perfect flower has both male (stamen) and female (ovary) reproductive organs on the same flower. (Compare with imperfect flowers.)

PERIANTH. The perianth (which means "around the anthers") is the sepals and petals of a flower.

PERICARP. The pericarp is the fruit wall that develops from the ovary wall. The pericarp is divided into the endocarp, mesocarp and exocarp.

PERICARPAL. The pericarpal is the upper part of the flower stem, the receptable, and the lower part of the pistil.

PERIOD. The period is the basic unit of geological time in which a single type of rock system is formed, lasting tens of millions of years.

PERMIAN EXTINCTION. The Permian extinction was the largest mass extinction that ever occurred on Earth, but its causes are unknown. It occurred at end of the Permian period, about 248 million years ago. It in, trilobites went extinct, as did 50% of all animal families, 95% of all marine species, and many trees. Groups that went extinct included: the fusulinid foraminifera, trilobites, rugose and tabulate corals, blastoids, acanthodians, placoderms, and pelycosaurs (like Dimetrodon). Groups that were substantially affected included: bryozoans, brachiopods, ammonoids, sharks, bony fish, crinoids, eurypterids, ostracodes, and echinoderms. This extinction was followed by the Triassic period of the Mesozoic Era.

PERMIAN PERIOD.

The Permian period (named after the Perm Province in northeast Russia where rocks from this period were first described) is known as "The Age of Amphibians" (280 to 248 million years ago), this is when Pangaea formed and Earth's atmosphere was oxygenated to modern levels. In the early Pemian, labyrinthodonts dominated the land. The Permian ended with the largest mass extinction and was followed by the Mesozoic Era.

PERMINERALIZATION. Permineralization is the process in which minerals are deposited into a bony fossil.

PETAL. A petal is one of the leafy structures that comprise a flower. Petals are often brightly-colored and have many different shapes. They are located between the sepals and the flower's reproductive organs.

PETIOLATE. Petiolate means having a petiole.

PETIOLE. A petiole is a leaf stalk. On a compound leaf, the petiole extends from the stem to the first set of leaflets. A leaf without a petiole is sessile.

PETRIFICATION. Petrification is the process in which an organic tissue turns to stone. The original materials are repaced by minerals.

PETRIFIED WOOD. Petrified wood is fossilized wood. Minerals seeped into buried logs, replacing the original tissues with rock.

PHANEROZOIC EON. The Phanerozoic (meaning "visible life") is the time in which life forms with skeletons or hard shells existed. It is the period from about 540 million years ago until the present.

PHENETICS. Phenetics is a method of attempting to classify biological organisms that does not use genetic or evolutionary information; it was invented by Sokal and Sneath in 1963. In a phenogram, organisms are grouped by superficial overall similarity. Phenetics was abandoned by most scientists in the 1980's because its classifications were arbitrary, mostly useless, and unstable. Paul Ehrlich was a proponent of this system.

PHLOEM. Phloem is plant tissue that conducts nutrients (food) through the plant. In woody-stemmed plants, the phloem is the inner layer of the bark. (Compare with xylem.)

PHOMA. (pronounced Phoma) Phoma is a genus of fungus that lives on a wide variety of plants (including sunflower, cabbage, banana, etc.). Phoma is characterized by large, black lesions on the stem and/or leaves of the infected plant. About 80 species of phoma have been described.

PHOTOSYNTHESIS .Photosynthesis is the process in which plants convert sunlight, water, and carbon dioxide into food (sugars and starches), oxygen and water. Chlorophyll or closely-related pigments (substances that color the plant) are essential to the photosynthetic process.

PHOTOTROPISM. Phototropism is the bending of a plant in response to sunlight. This reaction is caused by the growth hormone auxin that is contained in the stem.

PHYLLOCLADE. A phylloclade is a flattened stem that looks like a leaf.

PHYLLODE. A phyllode is a leaf that has an enlarged midrib and no blades

PHYLOGENY. Phylogeny is the evolutionary relationship between organisms. The phylogeny of an organism reflects the evolutionary branch that led up to the organism

PHYLUM. (plural plyla) In classification, a phylum is a group of related or similar organisms. A phylum contains one or more classes. A group of similar phyla forms a Kingdom. The plant phyla include: Ginkgophyta, Lycophyta (lower ferns like clubmosses), Pterophyta (ferns), Psilophyta (whisk ferns), Anthophyta (flowering plants), Gnetophyta,

Sphenophyta, Coniferophyta (conifers), Cycadophyta (cycads), Sphenophyta, and Bryophyta (mosses, liverworts, hornworts).

PINNATE COMPOUND LEAF.

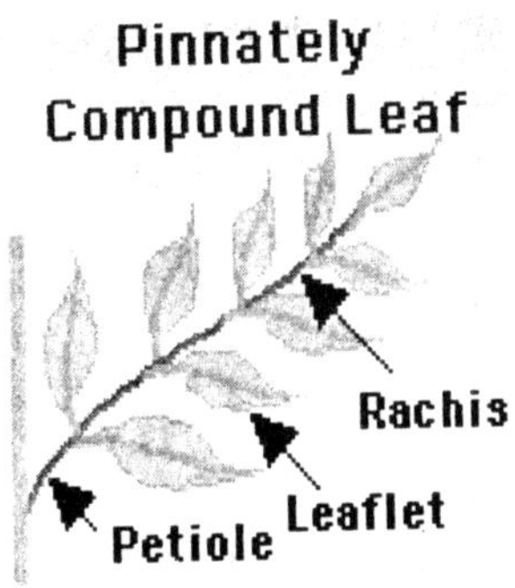

A pinnate compound leaf is made up of many small leaflets arranged in pairs on either side of a long central midrib (the rachis). There is often a single terminal leaflet at the end of the midrib.

PINNATELY LOBED LEAF.

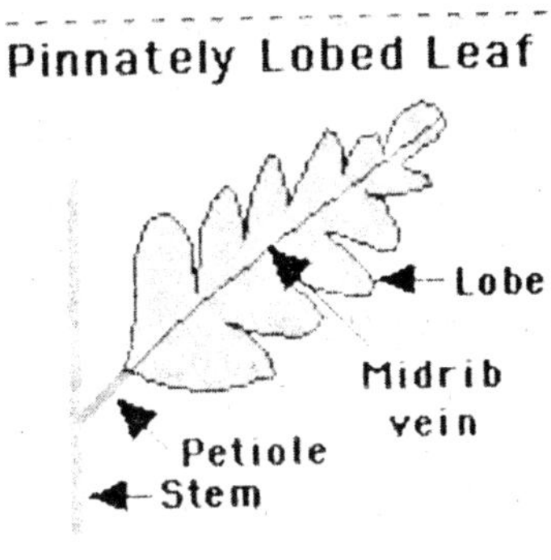

Pinnately lobed leaves have many lobes arranged along the midrib.

PINNATE VENATION A leaf with pinnate venation has its veins arranged in pairs coming from a main central midrib vein (the rachis).

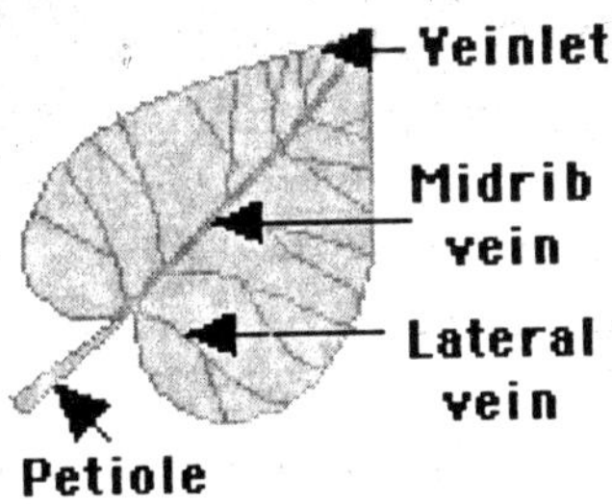

PIONEER. A pioneer is a plant that is the first (or among the first) to live in a new area

PISTIL. The pistil is the central set of female reproductive organs in a flower. The pistil is composed of one or more carpels and produces the ovule.

PISTILLATE. Pistillate flowers have a pistil or pistils.

PITH. The pith is parenchyma cells at the center of the primary stem of a dicot.

PLANT. A plant is a member of the kingdom Plantae, a living organism that undergoes photosynthesis.

PLANTAE. Plantae is a kingdom in the classificaton of life on Earth. Plantae (plants) is divided into non-vascular plants, vascular plants, and seed plants

PLANT SUCCESSION. Plant succession is the natural pattern of ecosystem growth and change over time for a particular environment. Plant life follows established patterns of growth and change after major distruptions, like fires, floods, agricultural damage, logging, etc. Generally, smaller, fast-growing herbaceous species and grasses grow first in an open field, followed in a few years by softwood tree seedlings and larger herbaceous species. As a young forest develops into a mature forest (30 to 70 years), an understory of smaller hardwood trees develops. The final stage is a climax hardwood forest (100 plus years).

PLASMA MEMBRANE. A plasma membrane (also called the cell membrane or plasmalemma) surrounds each cell and separates it from the environment. In pants, the cell membrane contains cellulose (a carbohydrate).

PLASMALEMMA. A plasmalemma (also called the cell membrane or plasma membrane) surrounds each cell and separates it from the environment. In pants, the cell membrane contains cellulose (a carbohydrate).

PLASMID. A plasmid is a circular loop of DNA found in prokaryotic cells (like those of bacteria). Eukaryotic cells prokaryotic cells (like plant and animal cells) do not have plasmids, they have chromosomes.

PLATE TECTONICS. Plate tectonics is the now-established theory that chunks of the Earth's crust (plates) float on the surface and change both position and size over time.

PLATYSPERMS. Platysperms (meaning "flat seeds") are plants with flattened ovules. Platysperms include gingkos, conifers, glossopterids, and cordaites.

PLEISTOCENE. The Pleistocene was an epoch of geologic time that lasted from 1.8 million to 11,000 years ago; it was a period of widespread glaciation and large Ice Age animals (caled Pleistocene Megafauna). The first humans (Homo sapiens) evolved during the Pleistocene. Mammoths, mastodons, saber-toothed cats, giant ground sloths, and other Ice Age mammals evolved. A mass extinction of large mammals and many birds occurred about 10,000 years ago, probably caused by Ice Ages.

PLESIOMORPHY. Plesiomorphy (meaning "old form") is a primitive character of a group.

PLUMULE. The plumule is the shoot of a plant embryo (in the seed before germination).

PNEUMATOPHORE. A pneumatophore is an air-transporting vessel in plants.

POD. A seed pod is an elongated, two-sided vessel that contain several fertilized seeds. It is a dehiscent fruit or pedicarp - the pod splits open when the seeds are mature. Beans and peas are some plants that have pods.

PODZOL. Podzol is a kind of soil that is very low in humus.

POLLEN. Pollen is the male reproductive cell of flowering plants and cone-bearing plants. Pollen grains are produced in the anther of a flower.

POLLENATION. Pollination is the transfer of pollen from the anther to the stigma.

POLLEN TUBE. After the male's pollen grains have landed on the stigma during fertilization, pollen tubes develop within the style. The pollen tubes transport the sperm from the grain to the ovum (where fertilization of the egg occurs and the seeds will develop).

POLYPHYLETIC. A polyphyletic group consists of organisms but not their common ancestors. This is an artificial group which is based primarily on physical characteristics rather than on evolutionary relationships. An example is "flying vertebrates" which includes birds, pteranodons and bats

POLLINATION. Pollination is the process in which the male's pollen fertilizes the female's ovule and creates a seed; the pollen is transferred from the male's anther to the female's stigma. Pollination is effected by the wind, insects, hummingbirds, etc.

POLLINIUM. A pollinium (or pollinia) is a mass of fused pollen grains that is produces by a single anther. This mass sticks together and during pollination is transported as a single unit. Orchids have polliniums.

PRECAMBRIAN. The Precambrian is the time 540 before the Cambrian period (before million years ago). It is the time from when the Earth formed until simple life-forms evolved.

PREHISTORIC. Prehistoric refers to the time before people began recording history in writing. This time varies from culture to culture.

PREY. An animal becomes prey when another animal hunts and kills it for food. Insects are the prey of the Venus fly trap.

PRICKLE. A prickle is a a sharp outgrowth of a plant's epidermis (outer skin). Prickles are easily broken off a plant. Roses have prickles.

PRICKLY PEAR CACTUS

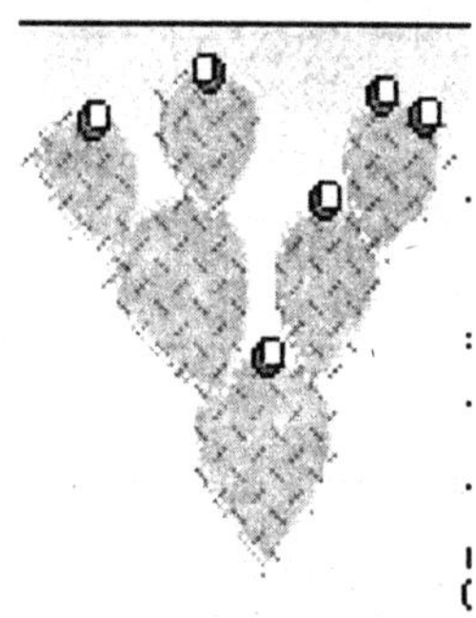

Prickly pear cacti (genus Opuntia, many species) are North American desert succulents that have flat, fleshy, leaf-shaped pads and large spines (modified leaves) growing from tubercles (small bumps on the pads). They have red, yellow, or purple flowers. Prickly pear cacti and their fruit are edible.

PRIMARY ROOT. The primary root is the first root of

a plant to develop in the germinating seed. The primary root develops from the radicle of the embryo. It is also called the taproot

PRIMARY VEGETATION. Primary vegetation is the plants life that has been in place since the area reached its climax state.

PRIMARY CONSUMER. A primary consumer is an herbivore (a plant-eater). Primary consumers occupy the second trophic level in the grazing food chain.

PRIMARY FOREST. A primary forest is a forest that has never been logged or disturbed.

PRIMARY GROWTH. Primary growth is plant growth that occurs at the tips (terminal bud) of the stem or the tip of the roots, at the apical meristems. In seed plants, primary growth produces .

PRODUCER. A producer (or aurotroph) is an organism that makes its own food from light energy (using photosynthesis), or chemical energy (using chemosynthesis). Most green plants, many protists (one-celled organisms like slime molds) and most bacteria are producers. Producers are the base of the food chain

PROTOPLAST. A protoplast is a plant or bacterial cell whose cell wall has been removed

PROTOZOA. Protozoans (meaning "first animals") are a phylum of primitive animals that include the following classes: Mastigophora (flagellates), Sarcodina (amoebas), Sporozoa (Parasites), and Ciliata (Ciliates).

PSAMMOPHILE. A psammophile (meaning "sand loving") is a plant that lives in sand or sandy soils. Some psammophytic plants include *Salix psammophila* (a willow), *Dypsis psammophila* (a palm) and *Cocolobo diversifolia* (the pigeon plum).

PSILOPHYTA. Psilophyta are a division of primitive plants that are also called wisk ferns (they are not true ferns). Psilophytes are seedless vascular plants with underground rhizomes. They live in tropical to sub-tropical areas and are terrestrial or epiphytic. There are only 2 genera (Psilotum and Tmesipteri), and only a few living species.

PTERIDOPHYTES. Pteridophytes are a group of primitive vascular plants that include Lycopods (club mosses), Sphenopsids (horsetails, shown left), and ferns (shown, right). These plants reproduce with spores that germinate only in moist areas; they also reproduce using rhizomes (underground stems). Pteridophytes evolved during the Devonian and were mostly low-growing during the Mesozoic Era. These fast-growing, resilient plants were a source of food for plant-eating dinosaurs that lived in moist areas.

PTERIDOSPERMS. Pteridosperms (Seed ferns) were primitive seed plants (not ferns at all) that lived in swampy areas from the Mississipian Epoch through the Mesozoic Era. They had woody stems studded with dried out leaf bases. The tops had fern-like fronds which bore seeds. Some seed ferns include Glossopteris (pictured above), Dicroidium, Caytonia, Denkania, and Lidgettonia.

PTEROPHYTA. Pterophyta (also called Pteridopsida) is a phylum of plants that includes some ferns, ancient vascular plants. There are about 11,000 living species of Pterophytes.

PULSE. A pulse refers to a leguminous plant that produces edible seeds or to the seeds themselves. Some pulses include peas, beands, and lentils.

PUMPKIN.

The pumpkin (*Cucurbita pepo*) is an annual vine from the New World.

Q

QUARTZ. Quartz is a common crystalline mineral.

QUATERNARY PERIOD. The Quaternary period, "The Age of Man" (1.8 million years ago to the present), is the most recent period of geological time.

QUEEN ANNE'S LACE. Queen Anne's lace, *Daucus carota*, is a wild flowering plant related to the carrot; it is also called the wild carrot. It has wide, flat flowers.

QUININE. Quinine is an alkaloid chemical that cures malaria, a tropical disease carried by mosquitoes. Quinine is found in the bark of the cinchona tree. South American natives have used cinchona bark to cure malaria for ages, and call the cinchona tree the "fever tree." In 1820, Pelletier and Caventou isolated a quinoline alkaloid from cinchona bark; they called it Quinine.

R

RACHIS.

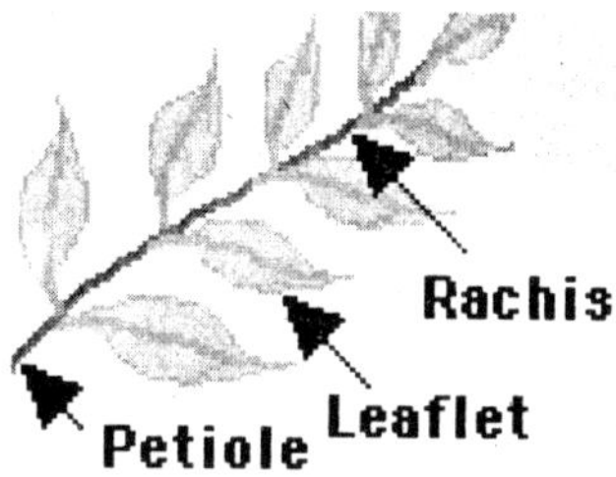

The rachis is the midrib of a leaf. It is usually continuous with the petiole and is often raised above the lamina (the leaf blade). On a compound leaf, the rachis extends from the first set of leaflets (where the petiole ends) to the end of the leaf.

RADIAL SYMMETRY. Radial symmetry is when a basic shape unit is repeated around a central point (for example, most flowers have radial symmetry)

RADICLE. The radicle is the lower part of a embryo's axis. The radicle develops into the primary root.

RADIOISOTOPE DATING. Radioisotope dating of igneous rock layers is used to find out how old the rock is (when the igneous rock formed). One way of dating fossils is by dating bracketing layers of igneous rock.

RAFFLESIA. The Rafflesia flower (*Rafflesia arnoldi*) is the world's largest flower. This giant bloom is found in rainforests of Indonesia. Locally, it is called the "corpse flower." Rafflesia gives off a putrid smell that reminds people of rotting meat (this odor attracts its pollenators, beetles and flies), hence its nickname. Rafflesia's enormous flower is about 3 feet (1 m) across and weighs about 20 pounds (9 kilograms). The flower takes about a year to develop, then it blooms for about a week before dying. The flower has five wide orange petals (with pale dots) surrounding a spiked cup. Rafflesia has no stem, no roots, and no leaves. The flower is supported by fungus-like tissue that lives in another plant - the Tetrastigma vine.

RAINFOREST. Rainforests are very dense, relatively warm, wet forests. They are havens for millions of plants and animals

RECEPTACLE. The receptacle is the terminal portion of the flower stalk.

REDWOOD. The redwood is the tallest tree, growing up to 370 feet (113 m) tall and living for over a thousand years. One redwood tree in California is 2,200 years old. The roots of this giant conifer are shallow, but spread sideways up to 250 feet (75 meters) from the trunk. The bark is deeply-furrowed, fibrous, thick [up to about 1 foot (30.5 cm) thick] and lacks resin. There are many species of redwood, including the giant coast redwood, *Sequoia sempervirens*.

REGRESSION. Regression is the exposure of continental land as the sea level decreases, usually caused by increasing polar ice and glacier formation. Another causes is the local uplift of the continental land.

RENIFORM. A reniform leaf is kidney-shaped.

REPAND. A repand leaf has a wavy margin.

RESIN. Resin is a substance secreted by some plants. Resin is often aromatic (having a strong odor); itt is insoluble in water but will dissolve in alcohol. Fossilized resin is called amber.

RESINIFEROUS. Resiniferous means producing resin. $C_6H_{12}O_6 + 6O_2 = 6CO_2 + 6H_2O$ **+ ATP**

RESPIRATION. Respiration is a process in which energy is generated in cells. In respiration, glucose (a type of sugar) and oxygen are converted into carbon dioxide, water and ATP (adenosine triphosphate). Respiration takes place in the mitochondrion of both plant and animal cells.

RETICULATE VENATION. A leaf with reticulate venation has its veins arranged in a pattern such that larger veins give rise to progressively smaller veins. The end branches of the veins define small areas called aeroles

RHIZOBIA. Rhizobia are bacteria that live symbiotically with plants (especially legumes), living on the plant's roots. The rhizobia fix nitrogen (from atmospheric nitrogen) for the plant and the rhizobia get energy from the plant.

RHIZOME. A rhizome is a thick, horizontal underground stem (not a root) of a plant, that grows close to the ground. Rhizomes have nodes and scale-like leaves; roots form on the lower surface and new shoots can form at nodes. Ferns, mosses, horsetails, ginger, irises, and some grasses have rhizomes

RIBOSOME. Ribosomes are small organelles composed of RNA-rich cytoplasmic granules that are sites of protein synthesis.

RIPARIAN FOREST. A riparian forest is situated by a river bank or other body of water.

RNA. RNA (short for ribonucleic acid) is a complex, organic, single-stranded molecule that is found in the nucleus and the cytoplasm of cells. RNA carries DNA's instructions for chemical synthesis (like protein formation) to the cell and guides the formation of these chemicals. There are many types of RNA that have different functions. The major types of RNA are messenger RNA (mRNA, which carries the information for protein synthesis from the chromosomal DNA to the ribosomes), transfer RNA (tRNA, which translates mRNA and bonds with amino acids to correctly form the desired proteins), and ribosomal RNA (rRNA, found in ribosomes).

ROOT.

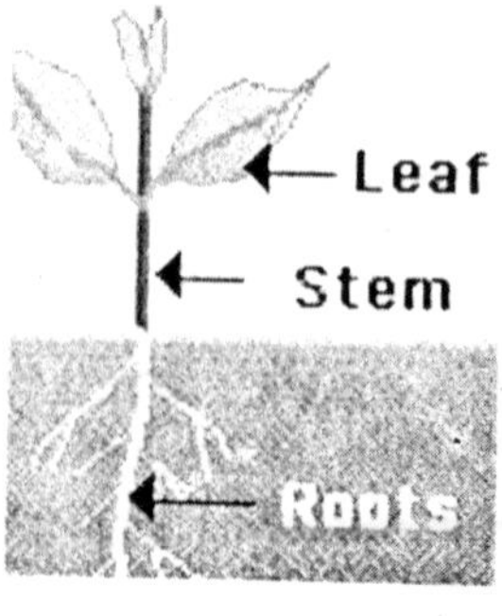

A root is a plant structure that obtains food and water from the soil, stores energy, and provides support for the plant. Most roots grow underground.

ROOT CAP. The root cap is a cap-shaped structure at the ends (tips) of the roots. It covers and protects the apical meristem (the actively growing region) of the root.

ROOT HAIRS. Root hairs are very thin, hair-like roots that are extensions of the root's epidermis. Root hairs have a large surface area and absorb water and minerals for the plant.

ROOT TIP. The root tip is the tip of the root and contains the root cap and the apical meristem (the actively growing region).

ROUGH ENDOPLASMIC RETICULUM. Rough ER is a vast system of interconnected, membranous, infolded and convoluted sacks that are located in the cell's cytoplasm (the ER is continuous with the outer nuclear membrane). Rough ER is covered with ribosomes that give it a rough appearance. Rough ER transport materials through the cell and produces proteins in sacks called cisternae (which are sent to the Golgi body, or inserted into the cell membrane).

RUBBER TREE. Rubber trees are large trees (belonging to the spurge family, family Euphorbiaceae) that live in tropical (warm) areas. These trees are tapped for their latex (from which rubber is made), which is produced in their bark layers (it is not the sap). The Pará rubber tree (*Hevea brasiliensis*) is native to South American rain forests, and grows to be over 100 ft (30 m) tall. In 1876, Henry Wickham brought seeds from the Para rubber tree (taken from the lower Amazon area of Brazil) to London, England. Seedlings were grown in London, and later sent to Ceylon and Singapore. The technique of tapping rubber trees for their latex was developed in southeast Asia (before that, the trees were cut down to extract the rubber). Commercial rubber production now takes place in Malaysia, Thailand, Indonesia, and Sri Lanka (but not significantly in South America).

RUGOSE. Rugose means ridged, rough, or wrinkled.

RUMINANT.

A ruminant is an animal that digests its food many times. This food is usually tough plant material like grasses. Ruminants include cows, sheep, antelopes, and camels.

RUNOFF. Runoff is water that drains or flows from the land into streams and rivers, eventually into seas. The water is generally from rain or snowpack melt.

S

SAGUARO CACTUS. (pronounced sah-WAH-row) The

saguaro cactus (*Carnegiea gigantea*) is a large succulent found in the Sonoran desert of North America. This cactus has accordian-like pleats and long spines. Photosynthesis takes place on its green, chlorophyll-containing trunk. The saguaro has a life span of over 200 years and it can grow to over 50 feet tall. This cactus produces white flowers and a deep red (edible) fruit. Each fruit produces about 2,000 reddish-black seeds.

SAMARA. A samara is a one-seeded fruit, winged, indehiscent fruit, like the seed of the maple and elm (which helicopter their way down from the tree). Samara is Latin for elm seed.

SAMAUMA TREE. The Samauma tree (*Eriodendron samauma*), sometimes called the "Queen of the Forest" or the silk-cotton tree, is a large, rainforest tree that grows to be over 50 m tall. It has an unusual lower trunk/roots that come off the main trunk in large, triangular planes. The soft-wood timber of this tree is pinkish-white. Many Samauma trees in the rainforest are being harvested to make inexpensive plywood.

SAP. Sap is a liquid that circulates within the sapwood of woody plants. Sap rises up from the roots. Sap contains water and minerals; in the spring it also contains sugars (and stimulates the growth of the tree).

SAPLING. A sapling is a small, young tree.

SAPROPHYTE. A saprophyte is a plant that obtains nutrition from dead and decaying plant or animal tissue. Most saprophytes do not produce chlorophyll, and therefore need another source of energy. Most fungi and a few flowering plants (like some orchids and Indian pipe) are saprophytic.

SAPWOOD. Sapwood is the outer layer of wood in a tree and contains living cells. Sap circulates within the sapwood of woody plants.

SAW GRASS. Saw grass, *Cladium jamaicense.*, is a plant that thrives in wet, warm, humid areas (swamps like the Florida Everglades). A member of the sedge family, saw grass has long, tough, sharp-toothed leaves.

SCAT. Scat means animal waste or droppings.

SCALE-LIKE LEAVES. Scale-like leaves are tiny, green leaves. On junipers, scale-like leaves overlap and cover the twigs.

SCLERENCHYMA. Sclerenchyma is a supportive and protective tissue found in plants. Sclerenchyma is composed of hard, thick, dry cells.

SCLEROPHYLL FOREST. A sclerophyll forest is one in which the crowns of the trees form a continuous canopy. The word scherophyll means "hard leaf" in Greek. Sclerophyll forests are often found in Australia - Eucalyptus trees often form a sclerophyll forest. There are wet (over 30 m tall) and dry (10 - 30 tall) sclerophyll forests

SCLEROPHYLLOUS PLANTS. Sclerophyllous plants are small plants that have hard, thickened leaves and have a relatively short distance along the stem between the leaves (short internodes). Sclerophyllous plants are often from dry areas. The word scherophyll means "hard leaf" in Greek.

SCYTHIAN EPOCH. (pronounced SY-thee-en EP-ock) The Scythian epoch was the early (lower) part of the Triassic period, about 248 million to 242 million years ago, the beginning of the Mesozoic Era.

SEASONS.

There are four seasons in the year: winter, spring, summer, and fall (also called autumn).

SEA WEED. Seaweed is a type of aquatic plant that obtains its energy via photosynthesis. Seaweed (like kelp) are not true plants. Seaweeds are usually green, brown, or red.

SECONDARY COMPOUND. A secondary compound is a chemical manufactured by a plant that protects it.

SECONDARY FOREST. A secondary forest (also known as a jungle) is a forest whose canopy trees have been logged (cut), causing lush growth on the forest floor.

SECONDARY GROWTH. Secondary growth is plant growth that does not occur at the tips of the stems or the tip of the roots. In seed plants, secondary growth produces bark and wood.

SEDIMENT. Sediment is any material deposited by wind or water, like rocks and sand.

SEDIMENTARY ROCK.

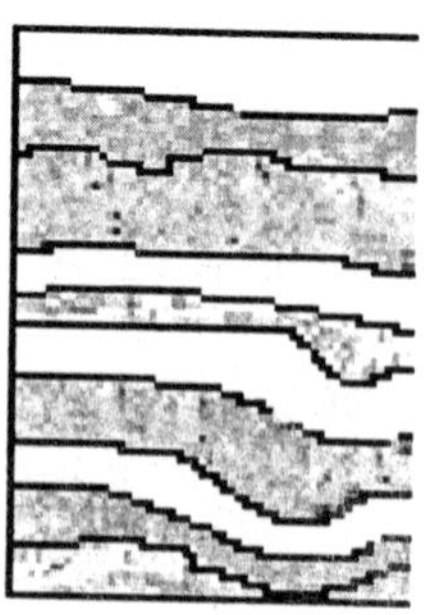

Sedimentary rock is rock that has formed from sediment. Most fossils are found in exposed sedimentary rock.

SEED. The seed is the reproductive unit of some plants.

SEED COAT. The seed coat is the outer, protective layer covering the seed. The seed coat is formed from the two integuments in the developing seed.

SEED DISPERSAL. Seeds are dispersed (spread) by many different methods, including floating on the wind (e.g., dandelions), floating in the water (e.g., coconut, sedge), hitching a ride on an animal (e.g., cranesbill), or being eaten (and then expelled) by a seed dispersing animal (e.g., many fruits).

SEED DISPERSER. A seed disperser is an animal that eats seeds (usually contained in fruit) but does not harm the seed. The seed is excreted in the stool, and the seed is spread away from the parent plant. Some seed dispersers include the kereru (a pigeon from New Zealand) and the cassowary (from Australia and New Guinea).

SEED FERNS.

Seed ferns (Pteridosperms) were primitive seed plants (not ferns at all) that lived in swampy areas from the Mississipian Epoch through the Mesozoic Era. They had woody stems studded with dried out leaf bases. The tops had fern-like fronds which bore seeds. Some seed ferns include Glossopteris (pictured above), Dicroidium, Caytonia, Denkania, and Lidgettonia.

SEED POD. A seed pod is an elongated, two-sided vessel that contain several fertilized seeds. It is a dehiscent fruit or pedicarp - the pod splits open when the seeds are mature. Beans and peas are some plants that have pods.

SEED PREDATOR. A seed predator is an animal that eats and destroy seeds instead of eating the fruit and leaving the seed or dispersing the seed in the stool. Parrots are seed predators. Seed predators limit the number of viable seeds.

SELF-POLLINATION. Self-pollination is the transfer of pollen from the anther to the stigma of the same flower or another flower on the same plant.

SENONIAN EPOCH. (pronounced cen-NOH-nee-an EP-ock) The Senonion epoch was the late (upper) part of the Cretaceous period, about 89 million to 65 million years ago, the end of the Mesozoic Era.

SEPAL. The sepals are small leaves located directly under a flower - they are the outermost part of a flower. Collectively, the sepals are called the calyx.

SERRATED. Serrated leaves have a jagged edge.

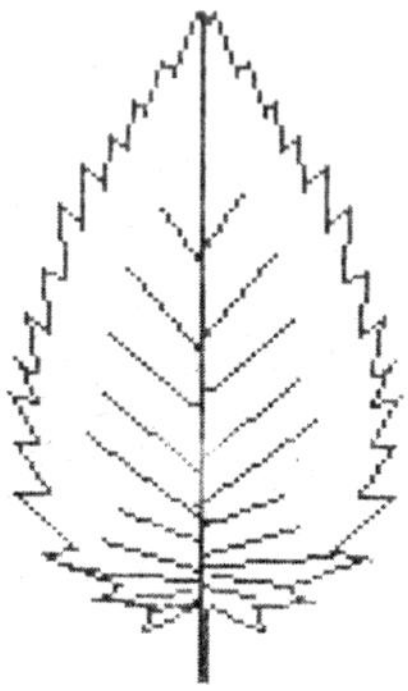

serrate

SESSILE. A leaf without a petiole (a leaf stalk) is sessile.

SEXUAL DIMORPHISM. Sexual dimorphism is characteristic of having two different forms, one for the males and another for the females of a species.

SHALE. Shale is a type of rock that is formed from clay that has been pressed into thin sheets.

SHIFTING CULTIVATION. Shifting cultivation is a type of farming in which fields are used for a few years, and are then left to grow in a wild state for many years. This allows the soil to recover and become rich and fertile again.

SHOCKED QUARTZ. Shocked quartz is quartz that has undergone deformation due to extreme pressure and heat. It has been found in the layer that marks the K-T boundary, lending credence to the Alvarez impact theory.

SHOOT. A shoot is new growth on part of a plant.

SHRUB LAYER. The shrub layer is the layer of the rainforest above the forest floor but under the canopy.

SIGNOR-LIPPS EFFECT. The Signor-Lipps Effect explains how a fossil record that appears to be a gradual extinction can actually represent a sudden extinction. If many organisms go extinct at the same time, the fossil record wouldn't necessarily represent the rarer species and the more common equally. The rarer species might disappear from the fossil record long before the time of extinction, simply due to chance.

SILT. Silt is fine dirt (soil or sand) that is carried by running water and deposited as sediment.

SILTATION. Siltation is the build-up of silt that is suspended in rivers or other bodies of water.

SILURIAN PERIOD. The Silurian Period lasted from 438 million to 408 million years ago. The first jawed fishes and uniramians (like insects, centipedes and millipedes) appeared during the Silurian (over 400 million years ago). The first vascular plants (plants with water-conducting tissue as compared with non-vascular plants like mosses) appeared on land (Cooksonia is the first known vascular plant). During this time, there were high seas worldwide until the end of the Silurian. Brachiopods, crinoids, corals, and eurypterids (sea scorpions) lived in the seas.

SIMPLE LEAF.

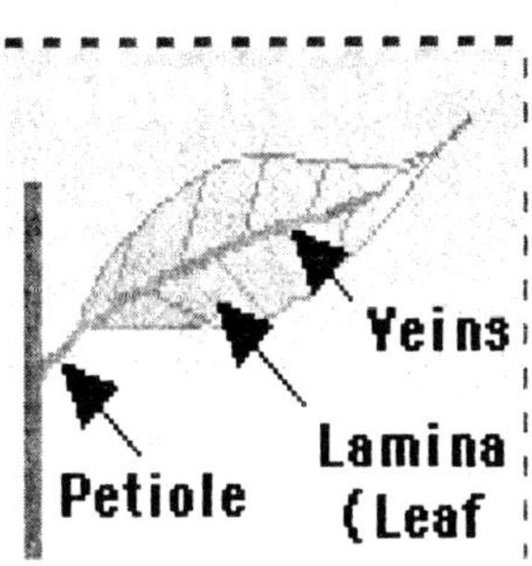

A simple leaf is a leaf with only one lamina for each petiole (that is, each leaf blade has one stem).

SIPHONOSTELE. A siphonostele is a type of stele that consists of a cylinder of vascular tissue (xylem and phloem) that surrounds the central pith tissue.

SLASH-AND-BURN FARMING. Slash-and-burn farming is a destructive type of agriculture in which the farmer burns down a new portion of the rainforest every few years in order to cultivate a crop.

SMOOTH ENDOPLASMIC RETICULUM. Smooth ER is a vast system of interconnected, membranous, infolded and convoluted tubes that are located in the cell's cytoplasm (the ER is continuous with the outer nuclear membrane). The space within the ER is called the ER lumen. Smooth ER transport materials through the cell. It contains enzymes and produces and digests lipids (fats) and membrane proteins; smooth ER buds off from rough ER, moving the newly-made proteins and lipids to the Golgi body, lysosomes, and membranes.

SOIL. Soil is a natural, constantly-changing substance that is made up of minerals, organic materials, and living organisms. plants grow in soil.

SOLAR RADIATION. Solar radiation is the heat and light that comes from the sun.

SP. Sp. is an abbreviation for "species." Sp. is often used when the genus is known, but the species is not.

SPECIATION. Speciation is the process in which a single species differentiates into two distinct species. One method by which this occurs is geographic isolation, in which two subpopulations of a single species are separated and no longer interbreed. Since the pressures of natural selection differ for the two groups, the two populations become more and more different from one other.

SPIKE. A spike is a flower stalk, an ear of grain (such as corn or wheat), or an inflorescence of unstalked flowers.

SPIKELET. A spikelet is a secondary spike found in grasses; it is a cluster of two or more flowers in the inflorescence.

SPINE. A spine is a sharp, modified leaf, scale, or stipule. Cacti have spines.

SPIROGYA. Spirogyra is a genus of multicellular green algae that lives in fresh water. Spirogyra cells are joined end-to-end, forming an unbranched, tube-like structure that contains one or two spirally-wound chloroplasts. Reproduction occurs asexually (as the filament breaks apart forming new strands) and sexually (two filaments meet and merge as the cell walls decay and a protoplast moves from one organism to the other, resulting in a zygote - this occurs under adverse conditions).

SPHENOPSIDS.

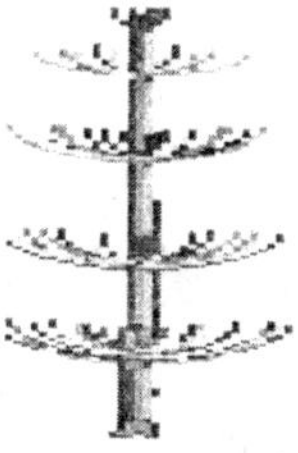

Sphenopsids (horsetails) are primitive, spore-bearing

plant with rhizomes. These fast-growing, resilient plants were common during the Mesozoic Era. The side branches are arranged in rings along the hollow stem. Horsetails date from the Devonian period 408-360 million years ago, but are still around today and are invasive weeds. Huge horsetails went extinct in the Permian mass extinction; smaller ones lived during the Mesozoic Era.

SPONGY MESOPHYLL. Spongy mesophyll is the layer below the palisade mesophyll; it has irregularly-shaped cells with many air spaces between the cells. These cells contain some chlorophyll. The spongy mesophyll cells communicate with the guard cells (stomata), causing them to open or close, depending on the concentration of gases.

SPORE. A spore is a single-celled reproductive unit of some organisms (cryptogams like mushrooms, ferns and mosses). Functionally, a spore is similar to a seed but it does not containan embryo). Spores are usually encapsulated by a rigid wall.

SPROUT. A sprout is a very young plant (newly germinated) or the new growth on a plant (a shoot).

STAMEN. The stamen is the male reproductive parts of a flower. It consists of the filament and the anther, which produces pollen.

STELE. A stele is the central cylinder of vascular bundles in stems and roots.

STEM. A stem is the axis of a plant; it may be above or below the ground.

STIGMA. The stigma is part of the pistil, the female reproductive tissue of a flower. The stigma receives the male pollen grains during fertilization.

STIPE. A stipe is a stem-like structure in some plants, like some kelp and fungi.

STIPULE. Stipules are small, paired appendages (sometimes leaf-life) that are found at the base of the petiole of leaves of many flowering plants.

STOLON. A stolon is an above-ground stem that has buds that sprout to form new shoots, forming a new, genetically-identical plant. Strawberry plants have stolons.

STOMA. A stoma (the plural is stomata) is a pore (or opening) in a plant's leaves. Most of the stoma are on the underside of the leaf. Guard cells open and close the stoma using turgor pressure, controlling the loss of water vapor and other gases from the plant.

STRATA. The strata (singular = stratum) are the different layers of a rainforest. Different animals and plants live in different parts of the rainforest. Scientists divide the rainforest into strata (zones) based on the living environment. Starting at the top, the strata are: emergents, canopy, understory, and forest floor.

STRATIGRAPHY. Stratigraphy is a method of dating fossils by observing how deeply a fossil is buried. Sedimentary rock layers (strata) are formed episodically as earth is deposited horizontally over time. Newer layers are formed on top of older layers, pressurizing them into rocks. Paleontologists can estimate the amount of time that has passed since the stratum containing the fossil was formed. Generally, deeper rocks and fossils are older than fossils found above them.

STRATOCLADISTICS. Stratocladistics is a method of classifying organisms based upon both cladistics (considering common ancestors with shared anatomical characteristics) together with stratiography (information from the fossil record which lets you know which animals lived earlier or later than others; older fossils are deeper than more recent fossils). In stratocladistics, cladograms are generated in which ancestors preceed their descendants. (see Science, 11th June, Vol 284, 1999)

STRIATE VENATION. Striate venation is a vein pattern found in monocots leaves. A leaf with striate venation has its veins arranged almost parallel to one another.

STROMA. The stroma is part of the chloroplasts in plant cells, located within the inner membrane of chloroplasts, between the grana.

STRONG SEASONALITY. Strong seasonality is when there is a big difference in temperature between the seasons (for example, a hot summer and a cold winter). Compare to low seasonality, in which the difference in temperatures between the seasons is small (with mild winters and summers).

STYLE. The style is part of the pistil, the female reproductive tissue of a flower. The style is a long tube on top of the ovary below the stigma. After the male's pollen grains have landed on the stigma during fertilization, pollen tubes develop within the style. The pollen tubes transport the sperm from the grain to the ovum (where fertilization of the egg occurs and the seeds will develop).

SUCCESSION, PLANT. Plant succession is the natural pattern of ecosystem growth and change over time for a particular environment. Plant life follows established patterns of growth and change after major distruptions, like fires, floods, agricultural damage, logging, etc. Generally, smaller, fast-growing herbaceous species and grasses grow first in an open field, followed in a few years by softwood tree seedlings and larger herbaceous species. As a young forest develops into a mature forest (30 to 70 years), an understory of smaller hardwood trees develops. The final stage is a climax hardwood forest (100 plus years).

SUCCULENT.

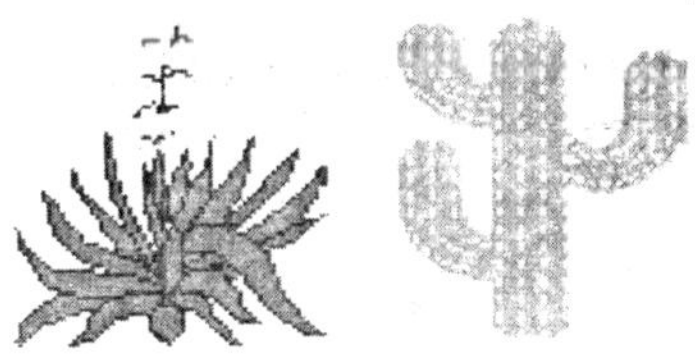

A succulent is a plant that has fleshy and juicy tissues, like cacti, sedums, aloes, and yuccas.

SUCKER ROOT. A sucker root is a root that emerges from the ground and sends up a shoot which supports the plant.

SUESS, EDUARD. Eduard Suess was an Austrian geologist who first realized that there had once been a land bridge between South America, Africa, India, Australia, and Antarctica. He named this large land mass Gondwanaland (named after a district in India where the fossil plant Glossopteris was found). This was the southern supercontinent formed after Pangaea broke up during the Jurassic period. He based his deductions upon the fossil fern Glossopteris, which is found throughout India, South America, southern Africa, Australia, and Antarctica.

SUSTAINABLE USE. Sustainable use is the judicious use of natural resources without destroying them.

SYMBIOSIS. Symbiosis is a situation in which two dissimilar organisms live together. There are many types of symbiosis, including mutualism (in which both organisms benefit), commensalism (in which one organism benefits and the other is not affected), or parasitism (in which one organism benefits at the other organism's expense). Symbiosis used to be defined as a situation in which two dissimilar organisms live together to the benefit of both - this is now called mutualism. The word symbiosis means "living together"" in Greek.

SYMMETRY. Symmetry across an axis (also called bilateral symmetry) is when one side of an object is the mirror image of its other half - i.e., one half has the same shape and size as the other half (for example, most leaves are bilaterally symmetrical). Radial symmetry is when a basic shape is duplicated around a central point (for example, most flowers have radial symmetry).

SYMPLESIOMORPHY. Symplesiomorphy (meaning "shared old form") is the persistence of ancestral (primitive) traits in different clades.

SYMPODIAL. Sympodial (meaning "with foot") is a type of branching growth in which the terminal bud dies or ends in an inflorescence, and growth (sympodial shoots) continues from lateral buds. What looks like the plant's main axis is actually a series of many lateral branches, each arising from the previous lateral branch. Some bamboos and orchids exhibits sympodial growth. Compare with monopodial.

SYNAPOMORPHY. Synapomorphy (meaning "shared form") is a derived (new) character shared by groups. A synapomorphy can be used to infer common ancestry.

T

TAP ROOT. The tap root is the main root of some plants; the tap root extends straight down under the plant with very little side branches (compare with fibrous root).

TAXON. A taxon is category in the classification of living organisms. The taxa (the plural of taxon) in the Linnean system are kingdom, phylum, class, order, family, genus, and species.

TAXONOMY. Taxonomy is the science of classifying organisms into groups by structure, origin, common ancestor, etc.

TEETH. Teeth are the jagged edges on some leaves, like those of thistles and hawthorne.

TEMPERATE RAIN-FOREST.

Temperate rainforests are very wet, ancient forests that rarely freeze or get very hot. Most of the world's temperate rainforests are in the Pacific Northwest of the USA.

TERMINAL BUD. The terminal bud is a bud located at the apex (tip) of the stem. This type of bud is the dominant bud, since it can cause all the lateral (side) buds below them to remain dormant at all times of the year. Terminal buds have special tissue, called apical meristem, cells that can divide indefinitely.

TERTIARY PERIOD. The Tertiary period lasted from 65 to 1.8 million years ago. It followed the Cretaceous period (the end of the Mesozoic Era) and the K-T extinction. Many mammals developed then, including primitive whales, rodents, pigs, cat, rhinos, etc.

TESTA. The testa is the seed coat. It covers the seed.

TETHYS SEA.

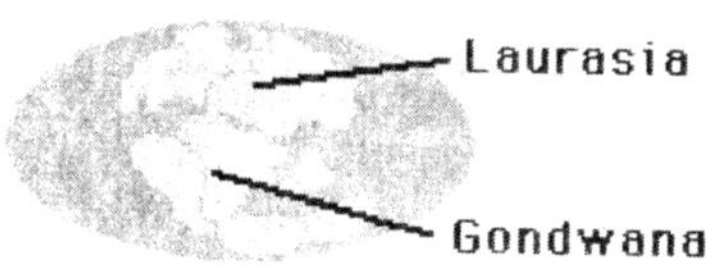

The Tethys sea was a shallow sea that existed during the early Mesozoic Era. It was the body of water that separated the landmass of Laurasia in the north from Gondwanaland in the south.

THEORY. A theory is a proposed explanation for a phenomenon.

THIGMOTROPISM. Thigmotropism is the directional bending or turning response of a plant upon contact with a solid surface or object; it is basically a sense of touch in plants. For example, the tendrils of vines are thigmotropic.

THORN. A thorn is a sharp, modified stem. Thorns have a stem-like vascular structure. The honey locust plant has thorns.

THYLAKOID DISK. Thylakoid disks are disk-shaped membrane structures in chloroplasts that contain chlorophyll. Chloroplasts are made up of stacks of thylakoid disks; a stack of thylakoid disks is called a granum. Photosynthesis (the production of ATP molecules from sunlight) takes place on thylakoid disks. Thylakoid means 'sac-like' in Greek.

TIPITI. A tipiti is a flexible braided cylindrical basket made of jacitara palm bark that is used by the Tupis of South America. The tipiti is used to remove the poison liquid from bitter cassava root. The cassava pulp is squeezed in the tipiti until the poisonous liquid is extracted. This process makes the cassava edible.

TOMATO. The tomato is scientifically considered to be a fruit (because the seeds of the plant are contained within the tomato).

TRACE FOSSILS. Also known as ichnofossils, these are fossilized footprints, nests, dung, gastroliths, etc., but not actual body parts. They record the movement and behavior of animals.

TRACHEOPHYTE. A tracheophyte is a plant with true vascular tissue, like xylem and phloem.

TRANSGRESSION. Transgression is the flooding of a continent as the sea level increases, usually caused by melting polar ice. Another cause is sea floor spreading and underwater volcanism, in which large amounts of underwater lava cause water to be displaced onto land.

TRANSPIRATION. Transpiration is the process in which plants lose water through pores in their leaves (these openings are called stomata). As water is lost from the plant, the plant takes up more water (and minerals) through its roots. The rate of transpiration varies as the conditions of the plant change and is controlled by the opening and closing the stomata.

TREE. A tree is a plant that produces wood (made by xylem cells). These tall plants grow taller each year.

TREE FERN. Tree ferns are tall vascular plants that live in warm climates. These ferns have a clump of fronds on top of a fibrous trunk.

TREE RINGS.

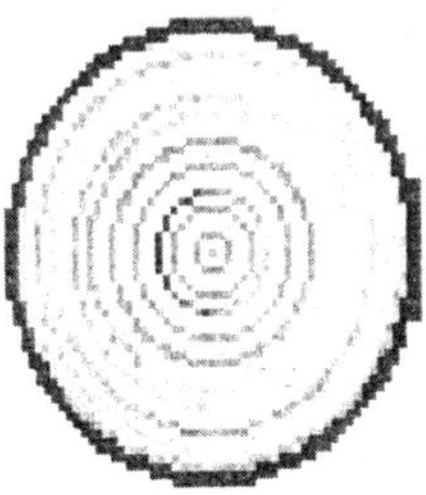

By counting the number of tree rings in a tree's trunk, you can tell how old the tree was.

TROPHIC LEVEL. A trophic level is a level of the grazing food chain. For example, plant-eaters are primary consumers; they occupy the second trophic level in the grazing food chain.

TRIASSIC PERIOD, (pronounced tri-ASS-ik PEER-ee-ud) Dinosaurs and mammals evolved during the Triassic period, 248 - 208 million years ago.

TROPICAL RAINFOREST. Tropical rainforests are warm, very wet forests that do not freeze (the temperature remains over 75°F=24°C) and do not get extremely hot. Tropical rainforests cover about 7% of the Earth's surface, in a band around the equator. They are also called equatorial rainforests.

TROPICS. The tropics are a 3,000 mile (4800 km) wide band around the equator, between the Tropic of Cancer (23.5° N latitude) and the Tropic of Capricorn (23.5° S latitude).

TRUNK. The trunk of a tree is the stem that supports the crown.

TSUNAMI.

(pronounced sue-NAHM-ee) A tsunami is a huge wave, caused by undersea earthquakes, volcanic eruptions, or, more rarely, by asteroid or meteoroid impact (as in the case of the K-T extinction).

TUBER. A tuber is a modified root that stores nutrients. Potatoes are tubers.

TUMBLEWEEDS. Tumbleweeds are annual plants that have a rounded shape; they dry our in the fall, and the stem beaks off near the ground. The ball-shaped, withered plant tumbles around in the wind (hence the name), scattering their seeds. There are many different types of tumbleweed - they live in the prairie and plains of the United States. One common type of tumbleweed is Russian thistle (*Salsola kali*).

TURGOR PRESSURE. Turgor pressure is the force that is exerted on a plant's cell wall by the water (in the cytoplasm) within the cell. The water presses against the cell wall from within, giving the plant cell rigidity, helping it keep its shape. Turgor pressure is the mechanism that changes the shape of the guard cells that open a plant's stomata, pores located on the underside of leaves.

TUSSOCK GRASS. Tussock grass (*Deschampsia cespitosa* or *Aira cespitosa*) is a tall variety of grass (20 to 36 inches tall) that is used as an ornamental grass. Tussock grass grows in clumps and has fine hair-like leaves. In summer it produces fine green to yellow inflourences.

TYPE SPECIES. A type species is the species of an organism from which a new genus is named. For example, Tyrannosaurus rex is the type species for the genus Tyrannosaurus.

TYPE SPECIMEN. A type specimen is the set of fossil remains of an organism from which a new species is named.

U

UNDERSTORY. The understory of a rainforest is a dark, cool environment under the leaves in the trees (the canopy) of a forest, but over the ground.

UNICELLULAR. Unicellular organisms consist of a single cells. (Compare with multicellular.).

URANIUM 235. Uranium 235 is a radioactive element; over time, it gradually changes into the element lead. It can be used to date very old rock layers.

V

VACUOLE. A vacuole is a large, membrane-bound space within a plant cell that is filled with fluid. Most plant cells have a single vacuole that takes up much of the cell. It helps maintain the shape of the cell.

VASCULAR PLANT.

A vascular plant has specialized pipelines that carry water and nutrients around the plant. Club mosses, ferns, horestails, gymnosperms, and flowering plants are vascular plants.

VASCULAR TISSUE. Vascular tissue is tissue that transports water and nutrients through a plant. Xylem and phloem are types of vascular tissue.

VEGETABLE.

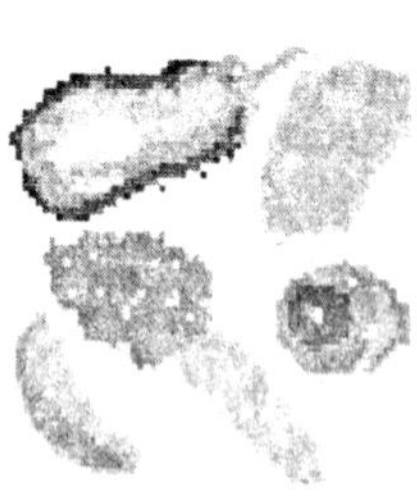

A vegetable is a plant whose stem, leaves, tubers, roots, bulbs, or flower is a food source for people. Some examples of vegetables include carrots, eggplant, potatoes, spinach, broccoli, onion, and asparagus.

VEGETATION. Vegetation is all of the plant life found in an area.

VEGETATIVE PROPAGATION. Vegetative propagation is a method of reproducing asexually; the offspring have the same genetic makeup as the parent.

VEGETATIVE STATE The vegetative state is the stage in a flowering plant's life cycle before the appearance of its fruiting structures.

VEIN. A vein is a vascular structure (xylem and phloem cells surrounded by the bundle sheath) in a leaf that provides supports for the leaf and transports both water and food. The veins on monocots are almost parallel to the margins of the leaf. The veins of dicots radiate from a central midrib.

VEINLET. A veinlet is a small vein. Veinlets are located toward the margins of the leaf.

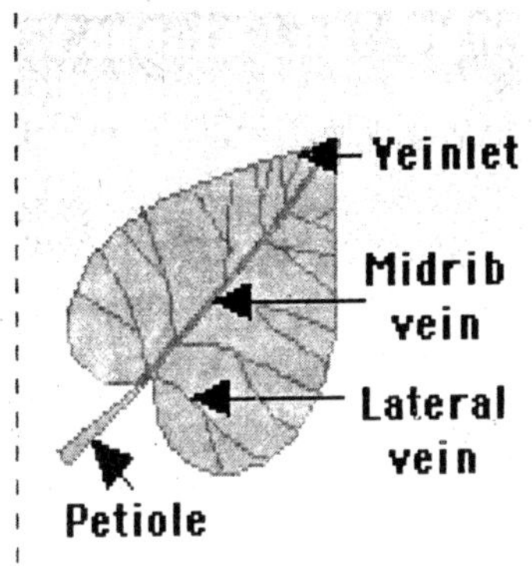

VENATION.

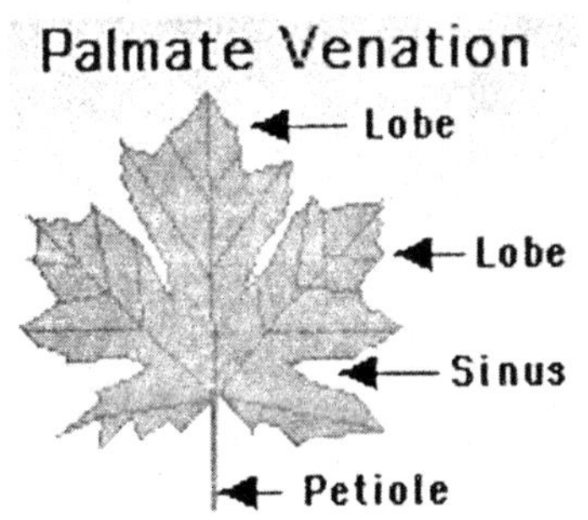

Venation is the arrangement of veins in a leaf. Some different cenation patterns include pinnate, palmate (illustrated above), and parallel.

VENDIAN. The Vendian or Ediacaran period was a geologic time period that lasted from 650 to 544 million years ago. The Vendian is when the earliest-known animals evolved. Vendian biota (Ediacara fauna), included soft-bodied multi-cellular animals, like sponges. During the Vendian, the continents had merged into a single supercontinent called Rodinia. The Vendian ended in a mass extinction.

VENUS FLYTRAP. The Venus flytrap is a plant that catches and digests bugs. Classification: Division Magnoliophyta (angioperms), Class Magnoliopsida (dicots), Subclass Dilleniidae, Order Nepenthales (insectivorous plants), Family Droseraceae (Sundews and Venus Flytrap).

VINE.

A vine is a plant that needs support as it grows. Some vines grow by twining around other objects for support (e.g., morning glory), some use tendrils (modified shoots) to attach to objects (e.g., peas and vetch), and others send out aerial roots, often with suckers (e.g., poison ivy).

VOLCANO.

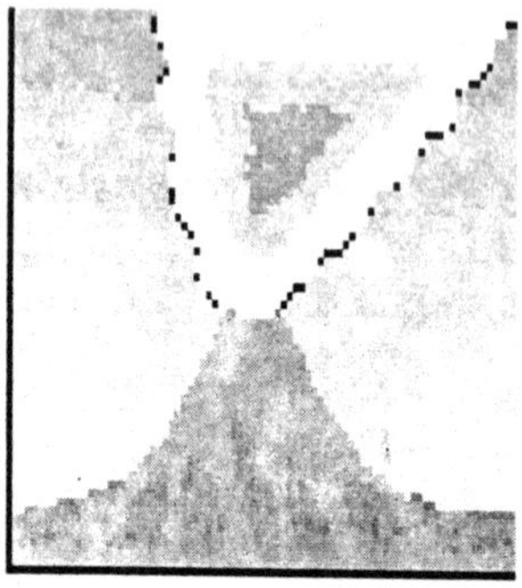

When a volcano erupts, it spews out lava and gases from deep inside the Earth. Radioisotope dating lava layers in rock sediment enables scientists to establish the date of fossils.

W

WATER CYCLE.

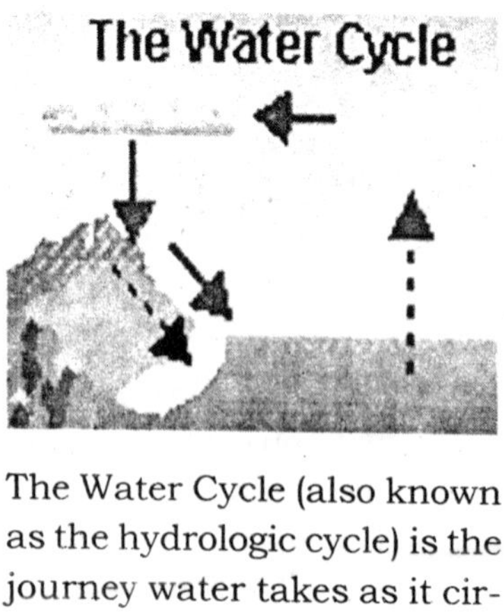

The Water Cycle (also known as the hydrologic cycle) is the journey water takes as it circulates from the Earth's surface to the sky and back again.

WAXY CUTICLE. The cuticle is the fatty or waxy outer layer of epidermal cells that are above ground.

WEALDEN FORMATION. The Wealden Formation is an area in southeastern England (including the Isle of Wight) that had yielded many fossils, iincluding dinosaurs. Paleontologist Gideon Mantell found Iguanodon there in 1825. The Wealden sedimentary rock dates from the Early Cretaceous period, when the area had fresh and brackish water (slightly salty) sediments.

WEGENER, ALFRED L.

Alfred Lothar Wegener (1880-1930), the German geologist and meteorologist, proposed the theory of continental drift in his 1915 book, *On the Origin of Continents and Oceans.* This theory states that parts of the Earth's crust slowly drift atop a liquid core. The fossil record overwhelmingly supports and gives credence to the theory of continental drift (and that of plate tectonics). He proposed the existence of the supercontinent Pangaea, and named it (Pangaea means "all the land" in Greek). Wegener froze to death while heading an expedition crossing the Greenland ice cap in 1930.

WELWITSCHIA.

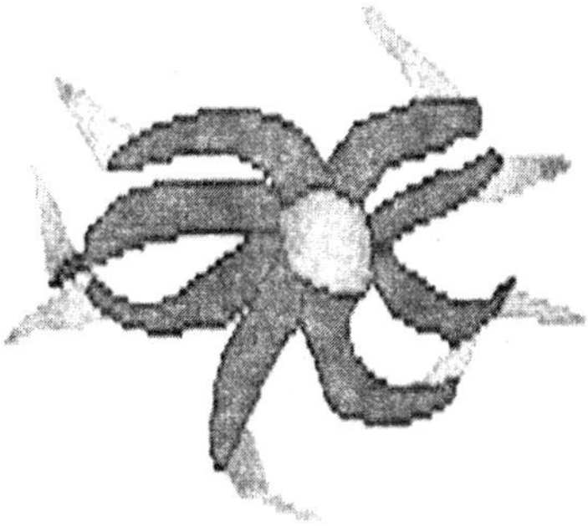

Welwitschia mirabilis is an unusual, long-lived, succulent gymnosperm. This plant has a low, two-lobed trunk, and two large, straplike leaves that sprawl along the ground. The plant only produces two leaves, but the wind splits them into many segments. It has a large, deep taproot. Welwitschia produces either male or female cones (but not both). Welwitschia lives in coastal desert areas, gravel beds, rocky canyon walls, and open flats. It derives its moisture from the morning fog. It is found from Namibia to southwestern Angola (in southwestern Africa). It has a life span of over 1,500 years. Welwitschia was first discovered by (and is named for) the Austrian botanist Dr. Friedrich Welwitsch in 1859. Classification: Division Pinophyta (Gymnosperms) , Subdivision Gneticae, Class Gnetopsida, Order Gnetales, Family Welwitschiaceae (Welwitschia)m Genus Welwitschia, Species *W. mirabilis.*

WILLIAMSONIA. Williamsonia is a fossil plant that thrived from the Triassic period through the Cretaceous period. Williamsonia was a bennettitalean a cycadeoidphyte, a primitive gymnosperm that resembled cycads but was not a cycad). Williamsonia had a long, thin, branching, woody trunk covered with spirals of broken-off leaf scars. It was up to 6.5 ft (2 m) tall.

WISCONSIN FAST PLANTS. Wisconsin Fast Plants are fast-growing plant varieties that were developed at the University of Wisconsin, Madison, Wisconsin, USA. Fast Plants complete their life cycles about 45 days after planting. Fast plants are used in classrooms and laboratories around the world for teaching and research. The first plants in space were Fast Plants.

WOOD.

Wood is the secondary xylem of gymnosperms and dicots, but the term wood is often applied to other xylem. Wood is used to make furniture, build houses, and make paper.

X

XERIC. Xeric environmental conditions are ones that are very dry. Compare with medium water conditions (mesic) and very wet conditions (hydric).

XEROPHYTE.

A xerophyte is a plant that is adapted to very dry conditions (like deserts and the emergent level of the rainforest). Succulents, bromeliads and cacti are xerophytes. Their adaptations include small or absent leaves, small, sunken stomata, and thick cuticles. (Compare with hydrophytes and mesophytes.)

XYLEM. (pronounced ZIE-lem) Xylem is a tissue in plant stems and roots. Xylem transports water and minerals upwards from the roots to the stem, via capillary action. Xylem is strong and also provides support to the plant. (Compare with phloem.)

Y

YUCCA .Yucca is a plant in the agave family that has stiff, sword-like leaves and clusters of white or purplish waxy flowers. This succulent plant is pollinated by the yucca moth. The Joshua tree (*Y. brevifolia*) is a type of yucca. Classification: Division Magnoliophyta, Class Liliopsida (monocots) , Subclass Lilidae, Order Liliales, Family Agavaceae (Agaves), Genus Yucca, many species.

Z

Zinnia.

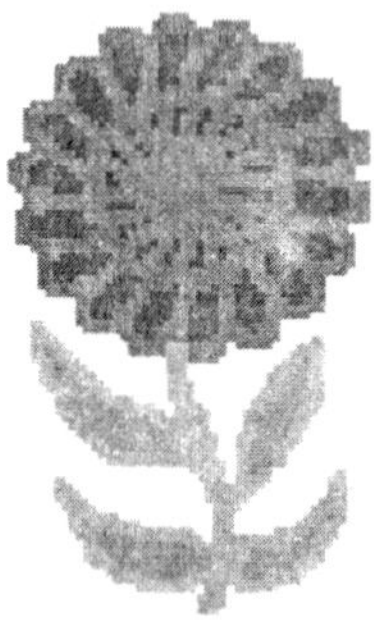

Zinnias are a genus of flowering annual plants that are native to Mexico. Zinnias are named for J. G. Zinn (1727-1759), a German botanist.

Zygote. A zygote is a diploid cell formed by the union of two gametes (it results from the union of an egg cell and a sperm cell).

ooo